Student Edition

S0-BOA-364

Eureka Math
Grade 1
Modules 5 & 6

Special thanks go to the Gordon A. Cain Center and to the Department of Mathematics at Louisiana State University for their support in the development of *Eureka Math*.

For a free *Eureka Math* Teacher Resource Pack, Parent Tip Sheets, and more please visit www.Eureka.tools

Published by the non-profit Great Minds™

Copyright © 2015 Great Minds. No part of this work may be reproduced, sold, or commercialized, in whole or in part, without written permission from Great Minds. Non-commercial use is licensed pursuant to a Creative Commons Attribution-NonCommercial-ShareAlike 4.0 license; for more information, go to http://greatminds.org/maps/math/copyright. "Great Minds" and "Eureka Math" are registered trademarks of Great Minds.

Printed in the U.S.A.

This book may be purchased from the publisher at eureka-math.org

10 9 8 7 6 5 4

v1.0 PAH

ISBN 978-1-63255-291-4

Name ___G_ Krishna Krishna___   Date _____

1. Circle the shapes that have 5 straight sides.

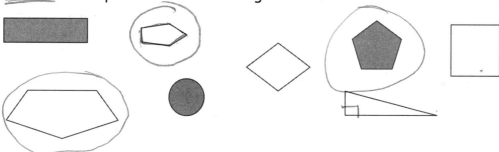

2. Circle the shapes that have no straight sides.

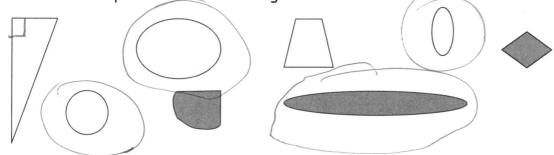

3. Circle the shapes where every corner is a square corner.

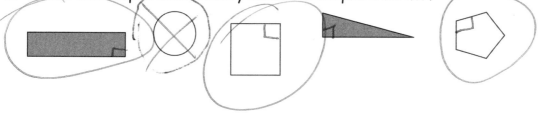

4.
a. Draw a shape that has 3 straight sides.	b. Draw another shape with 3 straight sides that is different from 4(a) and from the ones above.
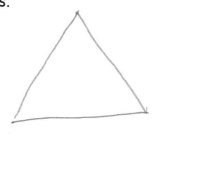	

EUREKA MATH

Lesson 1:   Classify shapes based on defining attributes using examples, variants, and non-examples.

1

©2015 Great Minds. eureka-math.org
G1-M5M6-SE-B4-1.3.1-01.2016

5. Which attributes, or characteristics, are the same for all of the shapes in Group A?

GROUP A

   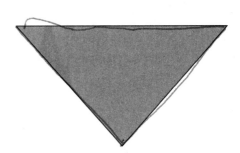

They all _____.

They all _____.

6. Circle the shape that best fits with Group A.

7. Draw 2 more shapes that would fit in Group A.	8. Draw 1 shape that would **not** fit in Group A.

Lesson 1:   Classify shapes based on defining attributes using examples, variants, and non-examples.

EUREKA MATH™

Name _Krishna_   Date _____

1. Circle the shapes that have 3 straight sides.

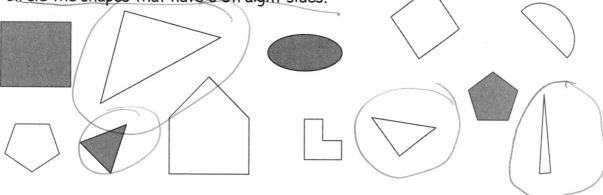

2. Circle the shapes that have no corners.

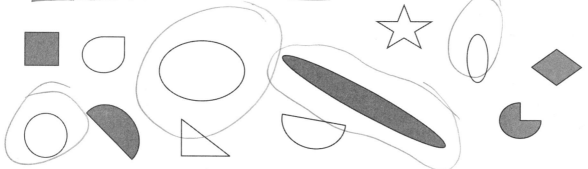

3. Circle the shapes that have only square corners.

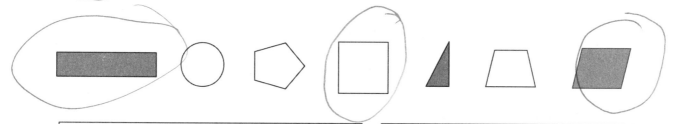

4.
a. Draw a shape that has 4 straight sides.	b. Draw another shape with 4 straight sides that is different from 4(a) and from the ones above.

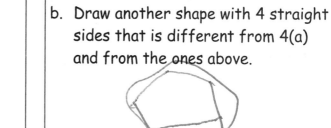

Lesson 1:   Classify shapes based on defining attributes using examples, variants, and non-examples.

3

©2015 Great Minds. eureka-math.org
G1-M5M6-SE-B4-1.3.1-01.2016

5. Which attributes, or characteristics, are the same for all of the shapes in Group A?

GROUP A

They all _____.

They all _____.

6. Circle the shape that best fits with Group A.

  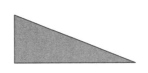 

7. Draw 2 more shapes that would fit in Group A.	8. Draw 1 shape that would **not** fit in Group A.

          Lesson 1:   Classify shapes based on defining attributes using examples, variants, and non-examples.   EUREKA
                                                                                                      MATH™

Name _Rhisma_____   Date _____

1. Use the key to color the shapes. Write how many of each shape are in the picture
    Whisper the name of the shape as you work.

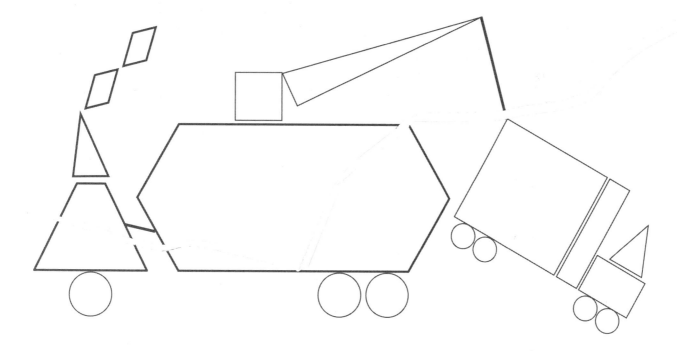

a. RED—4-sided shapes: _____   b. GREEN—3-sided shapes: _____

c. YELLOW—5-sided shapes: _____   d. BLACK—6-sided shapes: _____

e. BLUE—shapes with no corners: _____

2. Circle the shapes that are rectangles.

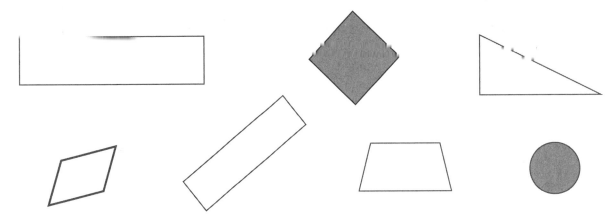

3. Is the shape a rectangle? Explain your thinking.

a.

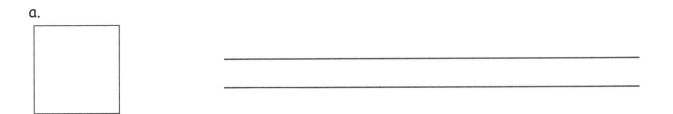

b.

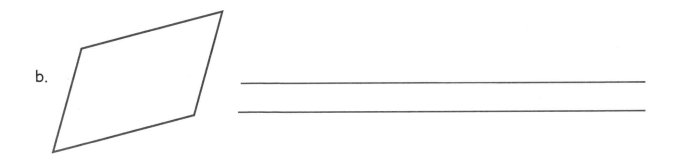

Lesson 2:   Find and name two-dimensional shapes including trapezoid, rhombus,
and a square as a special rectangle, based on defining attributes of
sides and corners.

EUREKA
MATH™

Name _____   Date _____

1. Color the shapes using the key. Write the number of shapes you colored on each line.

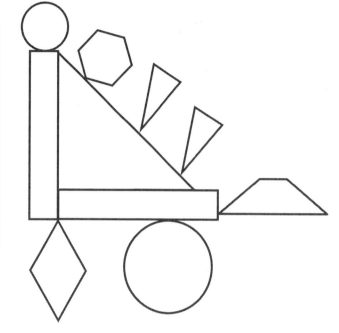

```
                    Key

RED  3 straight sides:      _____

BLUE  4 straight sides:     _____

GREEN  6 straight sides:    _____

YELLOW  0 straight sides:   _____
```

2.

   a. A **triangle** has ____ straight sides and ____ corners.

   b. I colored ____ triangles.

3.

   a. A **hexagon** has ____ straight sides and ____ corners.

   b. I colored ____ hexagon.

4.

   a. A **circle** has ____ straight sides and ____ corners.

   b. I colored ____ circles.

Lesson 2:   Find and name two-dimensional shapes including trapezoid, rhombus, and a square as a special rectangle, based on defining attributes of sides and corners.

©2015 Great Minds. eureka-math.org
G1-M5M6-SE-B4-1.3.1-01.2016

7

5.

   a. A **rhombus** has ____ straight sides that are equal in length and ____ corners.

   b. I colored ____ rhombus.

6. A **rectangle** is a closed shape with 4 straight sides and 4 square corners.

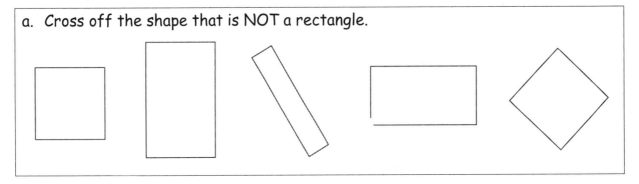

   a. Cross off the shape that is NOT a rectangle.

   b. Explain your thinking: _____

   _____

7. A **rhombus** is a closed shape with 4 straight sides of the same length.

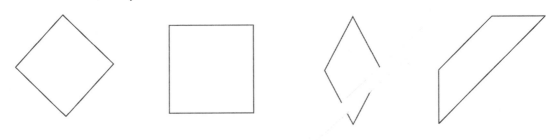

   a. Cross off the shape that is NOT a rhombus.

   b. Explain your thinking: _____

   _____

Lesson 2:   Find and name two-dimensional shapes including trapezoid, rhombus, and a square as a special rectangle, based on defining attributes of sides and corners.

EUREKA MATH™

Name _____   Date _____

1. On the first 4 objects, color one of the flat faces red. Match each 3-dimensional shape to its name.

a.    •

| Rectangular prism |

b. •

| Cone |

c. •

| Sphere |

d.    •

| Cylinder |

e. •

| Cube |

EUREKA MATH™

Lesson 3:   Find and name three-dimensional shapes including cone and
rectangular prism, based on defining attributes of faces and points.

9

2. Write the name of each object in the correct column.

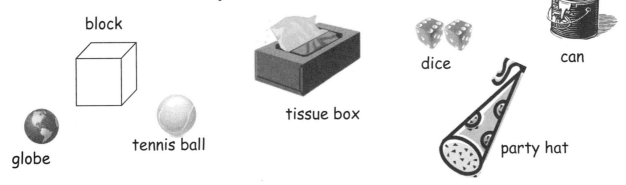

block

globe

tennis ball

tissue box

dice

can

party hat

Cubes	Spheres	Cones	Rectangular Prisms	Cylinders

3. Circle the attributes that describe ALL spheres.

have no straight sides

are round

can roll

can bounce

4. Circle the attributes that describe ALL cubes.

have square faces

are red

are hard

have 6 faces

Lesson 3:   Find and name three-dimensional shapes including cone and
                              rectangular prism, based on defining attributes of faces and points.

Name _____   Date _____

1. Go on a scavenger hunt for 3-dimensional shapes. Look for objects at home that would fit in the chart below. Try to find at least four objects for each shape.

Cube	Rectangular Prism	Cylinder	Sphere	Cone

Lesson 3:   Find and name three-dimensional shapes including cone and rectangular prism, based on defining attributes of faces and points.

11

©2015 Great Minds. eureka-math.org
G1-M5M6-SE-B4-1.3.1-01.2016

2. Choose one object from each column. Explain how you know that object belongs in that column. Use the word bank if needed.

Word Bank

faces	circle	square	roll	six
sides	rectangle	point	flat	

a. I put the _____ in the cube column because

_____.

b. I put the _____ in the cylinder column because

_____.

c. I put the _____ in the sphere column because

_____.

d. I put the _____ in the cone column because

_____.

e. I put the _____ in the rectangular prism column

because _____.

Lesson 3:   Find and name three-dimensional shapes including cone and rectangular prism, based on defining attributes of faces and points.

EUREKA MATH

Name _____   Date _____

Use pattern blocks to create the following shapes. Trace or draw to record your work.

1. Use 3 triangles to make 1 trapezoid.	2. Use 4 squares to make 1 larger square.
3. Use 6 triangles to make 1 hexagon.	4. Use 1 trapezoid, 1 rhombus, and 1 triangle to make 1 hexagon.

©2015 Great Minds. eureka-math.org
G1-M5M6-SE-B4-1.3.1-01.2016

5. Make a rectangle using the squares from the pattern blocks. Trace the squares to show the rectangle you made.

6. How many squares do you see in this rectangle?

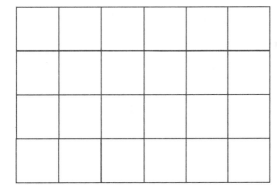

I can find _____ squares in this rectangle.

7. Use your pattern blocks to make a picture. Trace the shapes to show what you made. Tell a partner what shapes you used. Can you find any larger shapes within your picture?

EUREKA MATH™

Name _____   Date _____

Cut out the pattern block shapes from the bottom of the page. Color them to match the key, which is different from the pattern block colors in class. Trace or draw to show what you did.

Hexagon—red	Triangle—blue	Rhombus—yellow	Trapezoid—green

1. Use 3 triangles to make 1 trapezoid.	2. Use 3 triangles to make 1 trapezoid, and then add 1 trapezoid to make 1 hexagon.

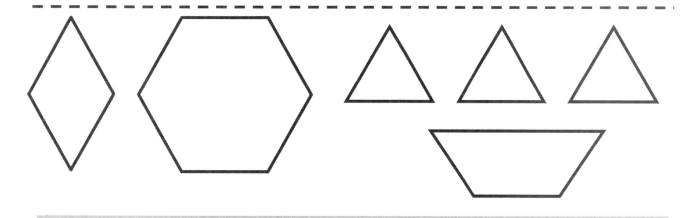

©2015 Great Minds. eureka-math.org
G1-M5M6-SE-B4-1.3.1-01.2016

This page intentionally left blank

3. How many squares do you see in this large square?

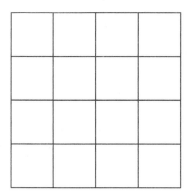

I can find _____ squares in this rectangle.

This page intentionally left blank

Name _____   Date _____

1.

    a. How many shapes were used to make this large square?

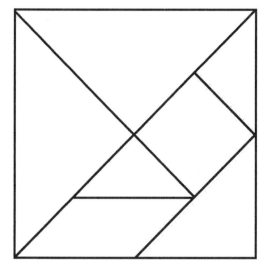

                                   There are _____
                                     shapes in this large square.

    b. What are the names of the 3 types of shapes used to make the large square?

_____   _____   _____

2. Use 2 of your tangram pieces to make a square. Which 2 pieces did you use? Draw or trace the pieces to show how you made the square.

3. Use 4 of your tangram pieces to make a trapezoid. Draw or trace the pieces to show the shapes you used.

©2015 Great Minds. eureka-math.org
G1-M5M6-SE-B4-1.3.1-01.2016

4. Use all 7 tangram pieces to complete the puzzle.

5. With a partner, make a bird or a flower using all of your pieces. Draw or trace to show the pieces you used on the back of your paper. Experiment to see what other objects you can make with your pieces. Draw or trace to show what you created on the back of your paper.

Lesson 5:   Compose a new shape from composite shapes.

EUREKA
MATH™

Name _____   Date _____

1. Cut out all of the tangram pieces from the separate piece of paper you brought home from school. It looks like this:

2. Tell a family member the name of each shape.

3. Follow the directions to make each shape below. Draw or trace to show the parts you used to make the shape.

   a. Use 2 tangram pieces to make 1 triangle.

   b. Use 1 square and 1 triangle to make 1 trapezoid.

   c. Use one more piece to change the trapezoid into a rectangle.

Lesson 5:   Compose a new shape from composite shapes.   **21**

©2015 Great Minds. eureka-math.org
G1-M5M6-SE-B4-1.3.1-01.2016

4. Make an animal with all of your pieces. Draw or trace to show the pieces you used.
    Label your drawing with the animal's name.

Lesson 5:   Compose a new shape from composite shapes.

EUREKA
MATH™

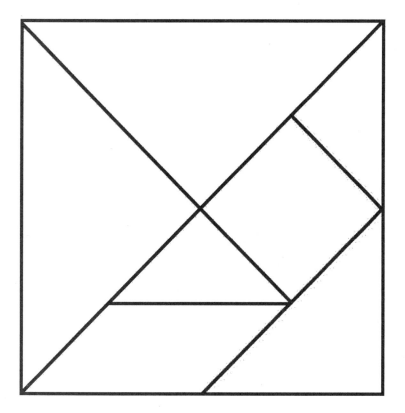

One tangram is to be used during class.
The other tangram is to be sent home with the homework.

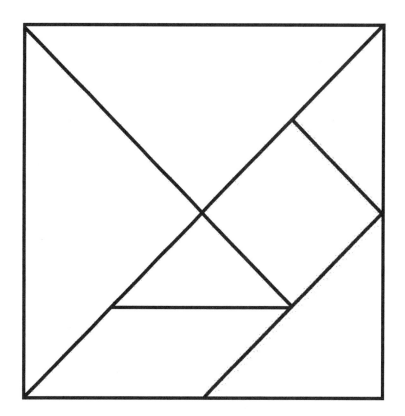

tangram

This page intentionally left blank

Name _____   Date _____

1. Work with your partner and another pair to build a structure with your 3-dimensional shapes. You can use as many of the pieces as you choose.

2. Complete the chart to record the number of each shape you used to make your structure.

Cubes	
Spheres	
Rectangular Prisms	
Cylinders	
Cones	

3. Which shape did you use on the bottom of your structure? Why?

4. Is there a shape you chose not to use? Why or why not?

Lesson 6:   Create a composite shape from three-dimensional shapes and describe the composite shape using shape names and positions.

25

©2015 Great Minds. eureka-math.org
G1-M5M6-SE-B4-1.3.1-01.2016

Name _____   Date _____

Use some 3-dimensional shapes to make another structure. The chart below gives you some ideas of objects you could find at home. You can use objects from the chart or other objects you may have at home.

Cube 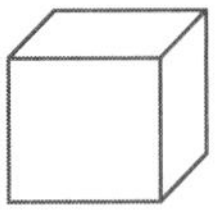	Rectangular prism 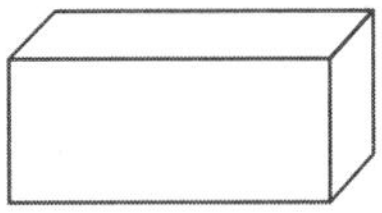	Cylinder 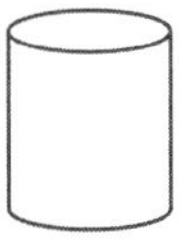	Sphere	Cone 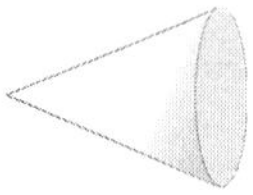
Block	Food box: Cereal, macaroni and cheese, spaghetti, cake mix, juice box	Food can: Soup, vegetables, tuna fish, peanut butter	Balls: Tennis ball, rubber band ball, basketball, soccer ball	Ice cream cone
Dice	Tissue box	Toilet paper or paper towel roll	Fruit: Orange, grapefruit, melon, plum, nectarine	Party hat
	Hardcover book	Glue stick	Marbles	Funnel
	DVD or video game box			

Ask someone at home to take a picture of your structure. If you are unable to take a picture, try to sketch your structure or write the directions on how to build your structure on the back of the paper.

Lesson 6:   Create a composite shape from three-dimensional shapes and describe the composite shape using shape names and positions.

EUREKA MATH™

Name _____ Date _____

1. Are the shapes divided into equal parts? Write **Y** for yes or **N** for no. If the shape has equal parts, write how many equal parts on the line. The first one has been done for you.

a.	b.	c.
Y ____ **2** ____	____ ____	____ ____
d.	e.	f.
____ ____	____ ____	____ ____
g.	h.	i.
____ ____	____ ____	____ ____
j.	k.	l.
____ ____	____ ____	____ ____
m. M	n. F	o. D
____ ____	____ ____	____ ____

EUREKA MATH

Lesson 7: Name and count shapes as parts of a whole, recognizing relative sizes of the parts.

27

©2015 Great Minds. eureka-math.org
G1-M5M6-SE-B4-1.3.1-01.2016

2. Write the number of equal parts in each shape.

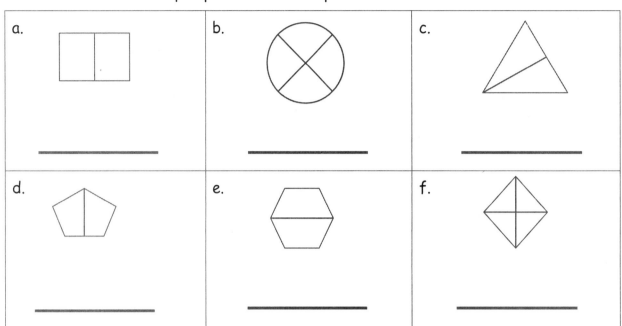

a. _____

b. _____

c. _____

d. _____

e. _____

f. _____

3. Draw one line to make this triangle into 2 equal triangles.

4. Draw one line to make this square into 2 equal parts.

5. Draw two lines to make this square into 4 equal squares.

Lesson 7:   Name and count shapes as parts of a whole, recognizing relative sizes of the parts.

EUREKA MATH

Name _____   Date _____

1. Are the shapes divided into equal parts? Write **Y** for yes or **N** for no. If the shape has equal parts, write how many equal parts there are on the line. The first one has been done for you.

a. **Y** _____ **2** _____	b. _____ _____	c. _____ _____
d. _____ _____	e. _____ _____	f. _____ _____
g. _____ _____	h. _____ _____	i. _____ _____
j. _____ _____	k. _____ _____	l. _____ _____
m. _____ _____	n. _____ _____	o. _____ _____

EUREKA MATH™

Lesson 7:   Name and count shapes as parts of a whole, recognizing relative sizes of the parts.

©2015 Great Minds. eureka-math.org
G1-M5M6-SE-B4-1.3.1-01.2016

29

2. Draw 1 line to make 2 equal parts. What smaller shapes did you make?

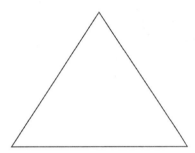

I made 2 _____.

3. Draw 2 lines to make 4 equal parts. What smaller shapes did you make?

I made 4 _____.

4. Draw lines to make 6 equal parts. What smaller shapes did you make?

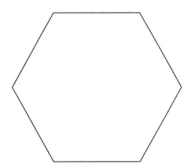

I made 6 _____.

Lesson 7: Name and count shapes as parts of a whole, recognizing relative sizes of the parts.

EUREKA MATH

Name _____　　　Date _____

1.　Are the shapes divided into halves?　Write yes or no.

a.	b. 	c. 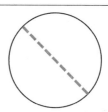
d.	e.	f. 

2.　Are the shapes divided into quarters?　Write yes or no.

a. 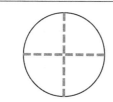	b. 	c. 
d. 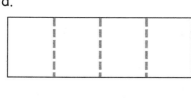	e. 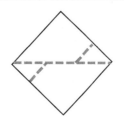	f. 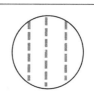

3. Color half of each shape.

a.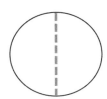

b.

c.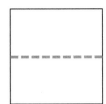

d.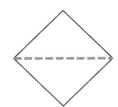

e.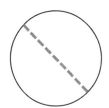

f.

4. Color 1 fourth of each shape.

a.

b.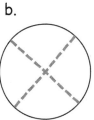

c.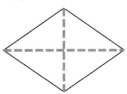

d.

e.

Lesson 8:   Partition shapes and identify halves and quarters of circles and rectangles.

EUREKA MATH™

Name _____   Date _____

1. Circle the correct word(s) to tell how each shape is divided.

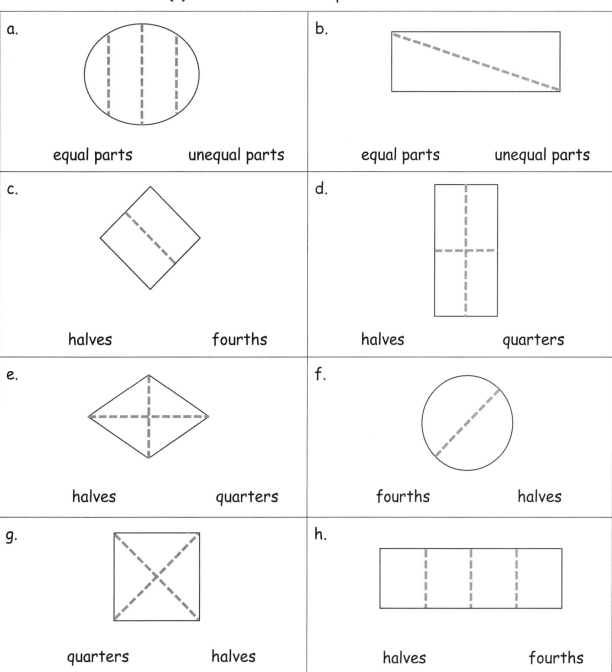

a.

equal parts   unequal parts

b.

equal parts   unequal parts

c.

halves   fourths

d.

halves   quarters

e.

halves   quarters

f.

fourths   halves

g.

quarters   halves

h.

halves   fourths

Lesson 8:   Partition shapes and identify halves and quarters of circles and rectangles.

33

©2015 Great Minds. eureka-math.org
G1-M5M6-SE-B4-1.3.1-01.2016

2. What part of the shape is shaded? Circle the correct answer.

a.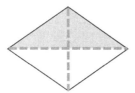

    1 half   1 quarter

b.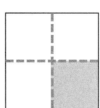

    1 half   1 quarter

c.

    1 half   1 quarter

d.

    1 half   1 quarter

3. Color 1 quarter of each shape.

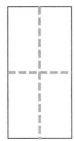 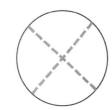

4. Color 1 half of each shape.

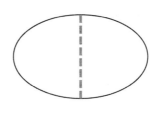 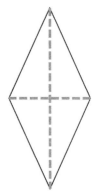

Lesson 8:   Partition shapes and identify halves and quarters of circles and rectangles.

EUREKA MATH

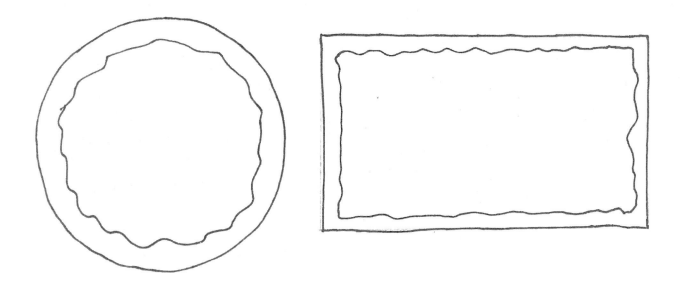

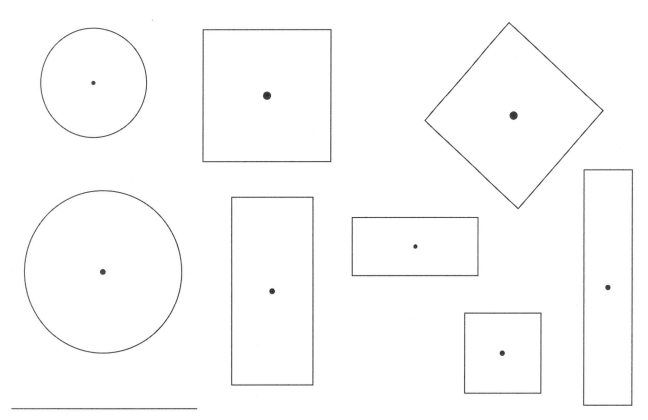

circles and rectangles

Lesson 8:   Partition shapes and identify halves and quarters of circles and rectangles.

35

This page intentionally left blank

Name _____   Date _____

Label the shaded part of each picture as one half of the shape or one quarter of the shape.

1.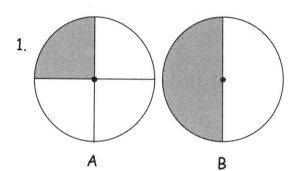

            A   B

Which shape has been cut into more equal parts? _____

Which shape has larger equal parts? ____

Which shape has smaller equal parts? ____

2.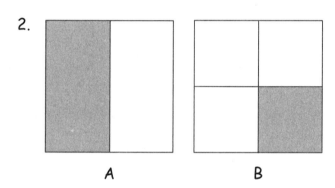

       A   B

Which shape has been cut into more equal parts? _____

Which shape has larger equal parts? ____

Which shape has smaller equal parts? ____

3. Circle the shape that has a larger shaded part. Circle the phrase that makes the sentence true.

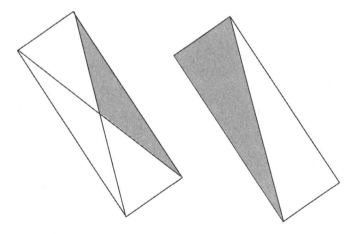

The larger shaded part is

(one half of / one quarter of)

the whole shape.

Color part of the shape to match its label.

Circle the phrase that would make the statement true.

4.

One half of the circle

is larger than

is smaller than

is the same size as

one fourth of the circle.

5.

One quarter of the rectangle

is larger than

is smaller than

is the same size as

one half of the rectangle.

6.

One quarter of the square

is larger than

is smaller than

is the same size as

one fourth of the square.

Lesson 9: Partition shapes and identify halves and quarters of circles and rectangles.

Name _____   Date _____

1. Label the shaded part of each picture as one half of the shape or one quarter of
   the shape.

A

Which picture has been cut into more equal parts? _____

Which picture has larger equal parts? ____

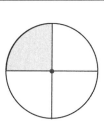

B

Which picture has smaller equal parts? ____

2. Write whether the shaded part of each shape is a half or a quarter.

a.	b.
 _____ _____	 _____ _____
c.	d.
 _____ _____	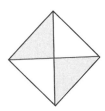 _____ _____

EUREKA
MATH™

Lesson 9:   Partition shapes and identify halves and quarters of circles and
rectangles.

39

3. Color part of the shape to match its label. Circle the phrase that would make the statement true.

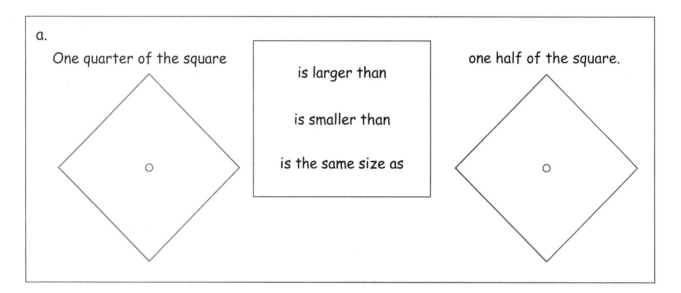

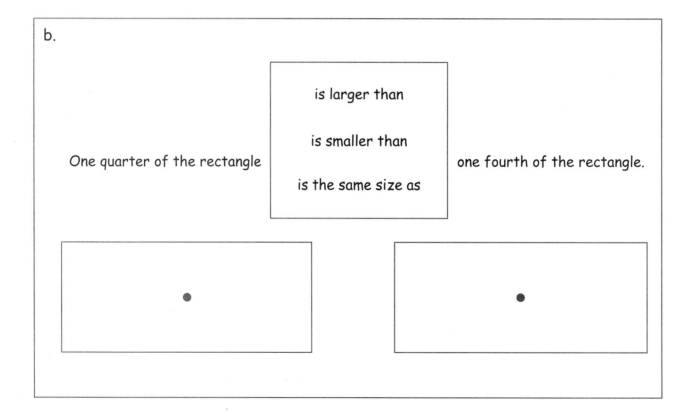

        Lesson 9:   Partition shapes and identify halves and quarters of circles and rectangles.

©2015 Great Minds. eureka-math.org
G1-M5M6-SE-B4-1.3.1-01.2016

EUREKA MATH™

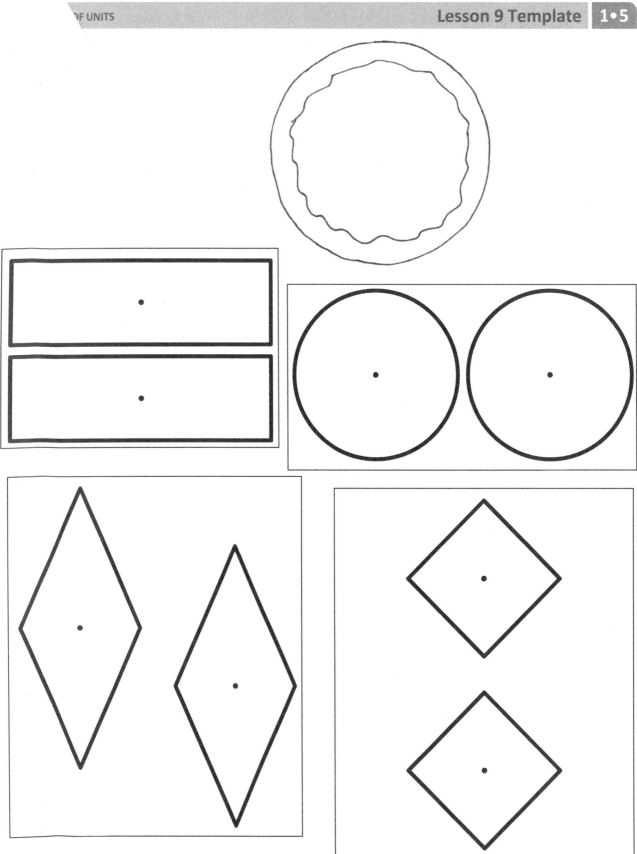

pairs of shapes

 EUREKA MATH™

Lesson 9:   Partition shapes and identify halves and quarters of circles and
rectangles.

This page intentionally left blank

Name _____   Date _____

1. Match the clocks that show the same time.

a.	b.	c.	d.

2. Put the hour hand on this clock so that the clock reads 3 o'clock.

EUREKA MATH

Lesson 10:   Construct a paper clock by partitioning a circle and tell time to the hour.

43

©2015 Great Minds. eureka-math.org
G1-M5M6-SE-B4-1.3.1-01.2016

3. Write the time shown on each clock.

a. _____ : _____

b. _____ o'clock

c. _____ o'clock

d. _____ o'clock

e. _____ : _____

f. _____ o'clock

g. _____ : _____

h. _____ o'clock

i. _____ : _____

j. _____ o'clock

k. _____ : _____

l. _____ o'clock

m. _____

n. _____

o. _____

Lesson 10:   Construct a paper clock by partitioning a circle and tell time to the
              hour.

EUREKA
MATH™

Name _____   Date _____

1. Match each clock to the time it shows.

a.

b.

| 4 o'clock |

c.

| 7 o'clock |

d.

| 11 o'clock |

e.

| 10 o'clock |

| 3 o'clock |

f.

| 2 o'clock |

Lesson 10:   Construct a paper clock by partitioning a circle and tell time to the hour.

45

EUREKA MATH

2. Put the hour hand on the clock so that the clock matches the time. Then, write the time on the line.

a.

| 6 o'clock |

6 : 00

_____

b.

| 9 o'clock |

_____

c.

| 12 o'clock |

_____

d.

| 7 o'clock |

_____

e.

| 1 o'clock |

_____

Lesson 10: Construct a paper clock by partitioning a circle and tell time to the hour.

EUREKA MATH™

Name _____   Date _____

1. Match the clocks to the times on the right.

a.    ●

b.    ●

c.    ●

● ┌─────────────────────┐
  │ Half past 5 o'clock │
  └─────────────────────┘

● ┌─────────────────────┐
  │   12:30   │
  └─────────────────────┘

● ┌─────────────────────┐
  │   2:30   │
  └─────────────────────┘

● ┌─────────────────────┐
  │ Five thirty   │
  └─────────────────────┘

● ┌─────────────────────┐
  │ Half past 12 o'clock│
  └─────────────────────┘

● ┌─────────────────────┐
  │ Two thirty   │
  └─────────────────────┘

2. Draw the minute hand so the clock shows the time written above it.

a. 7 o'clock   b. 8 o'clock   c. 7:30

d. 1:30   e. 2:30   f. 2 o'clock

EUREKA
MATH™

3. Write the time shown on each clock. Complete problems like the first two examples.

a.	b.	c.
3:30	five thirty	_____
d.	e.	f.
_____	_____	_____
g.	h.	i.
_____	_____	_____
j.	k.	l.
_____	_____	_____

4. Circle the clock that shows half past 12 o'clock.

a.   b.   c.

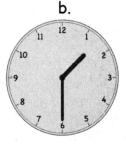
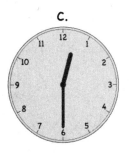

Lesson 11:   Recognize halves within a circular clock face and tell time to the half
hour.

EUREKA
MATH™

Name _____   Date _____

Circle the correct clock.

1. Half past 2 o'clock

    a.   b.   c.

 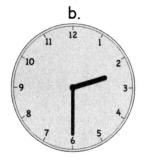 

2. Half past 10 o'clock

    a.   b.   c.

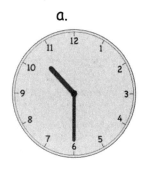  

3. 6 o'clock

    a.   b.   c.

4. Half past 8 o'clock

    a.   b.   c.

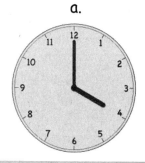 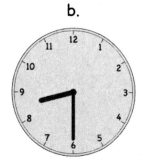 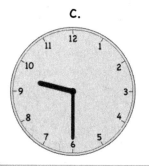

Write the time shown on each clock to tell about Lee's day.

5.  Lee wakes up at _____.	6.  He takes the bus to school at _____.
7.  He has math at _____.	8.  He eats lunch at _____.
9.  He has basketball practice at _____.	10.  He does his homework at _____.
11.  He eats dinner at _____.	12.  He goes to bed at _____.

Lesson 11:   Recognize halves within a circular clock face and tell time to the half hour.

EUREKA MATH™

Name _____   Date _____

Fill in the blanks.

1.

A   B

Clock _____ shows half past eleven.

2.

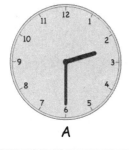 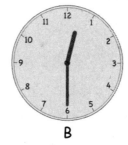

A   B

Clock _____ shows half past two.

3.

A   B

Clock _____ shows 6 o'clock.

4.

 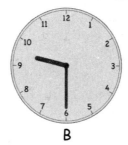

A   B

Clock _____ shows 9:30.

5.

A   B

Clock _____ shows half past six.

EUREKA MATH™

Lesson 12:   Recognize halves within a circular clock face and tell time to the half hour.

51

6. Match the clocks.

a.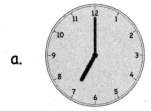

b.

c.

d.

half past 7

half past 1

7 o'clock

half past 5

 7:30

 7:00

 5:30

 1:30

7. Draw the minute and hour hands on the clocks.

a. 3:30

b. 8:30

c. 11:00

d. 6:00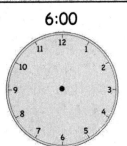

e. 4:30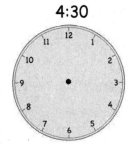

f. 12:30

Lesson 12:   Recognize halves within a circular clock face and tell time to the half hour.

EUREKA MATH

Name _____ Date _____

Write the time shown on the clock, or draw the missing hand(s) on the clock.

1. 10 o'clock	2. half past 10 o'clock
3. 8 o'clock	4. _____
5. 3 o'clock	6 half past 3 o'clock
7. _____	8. half past 6 o'clock
9. half past 9 o'clock	10. 4 o'clock

Lesson 12:   Recognize halves within a circular clock face and tell time to the half hour.

53

11. Match the pictures with the clocks.

a.

| Soccer practice |
| 3:30 |

b.

| Brush teeth |
| 7:30 |

c.

| Wash dishes |
| 6:00 |

d.

| Eat dinner |
| 5:30 |

e.

| Take bus home |
| 4:30 |

f.

| Homework |
| half past 6 o'clock |

Lesson 12: Recognize halves within a circular clock face and tell time to the half hour.

EUREKA MATH™

Name _____ Date _____

Circle the correct clock. Write the times for the other two clocks on the lines.

1. Circle the clock that shows half past 1 o'clock.

a. 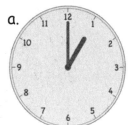 b. 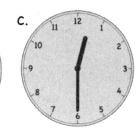 c.

_____

_____

2. Circle the clock that shows 7 o'clock.

a.  b. 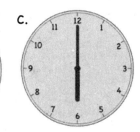 c.

_____

_____

3. Circle the clock that shows half past 10 o'clock.

a.  b. 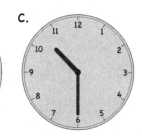 c.

_____

_____

4. What time is it? Write the times on the lines.

a. 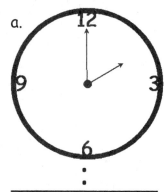 b.  c.

____ : ____   ____ : ____   ____ : ____

EUREKA MATH

Lesson 13:   Recognize halves within a circular clock face and tell time to the half hour.

55

5. Draw the minute and hour hands on the clocks.

a. 1:00
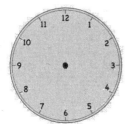

b. 1:30

c. 2:00
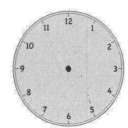

d. 6:30

e. 7:30

f. 8:30
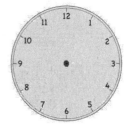

g. 10:00
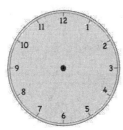

h. 11:00
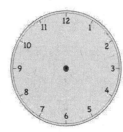

i. 12:00

j. 9:30
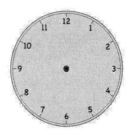

k. 3:00

l. 5:30
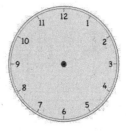

Lesson 13:   Recognize halves within a circular clock face and tell time to the half hour.

EUREKA MATH™

©2015 Great Minds. eureka-math.org
G1-M5M6-SE-B4-1.3.1-01.2016

Name _____   Date _____

Fill in the blanks.

1.

Clock _____ shows half past three.

2.

Clock _____ shows half past twelve.

3.

Clock _____ shows eleven o'clock.

4.

Clock _____ shows 8:30.

5.

Clock _____ shows 5:00.

Lesson 13:   Recognize halves within a circular clock face and tell time to the half
              hour.

57

6. Write the time on the line under the clock.

a.	b.	c.
_____	_____	_____

d.	e.	f.

_____	_____	_____

g.	h.	i.

_____	_____	_____

7. Put a check (✓) next to the clock(s) that show 4 o'clock.

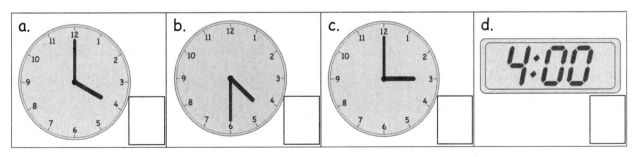

EUREKA
MATH™

©2015 Great Minds. eureka-math.org
G1-M5M6-SE-B4-1.3.1-01.2016

clock images

Lesson 13:   Recognize halves within a circular clock face and tell time to the half hour.

59

EUREKA
MATH™

This page intentionally left blank

Eureka Math
Grade 1
Module 6

Special thanks go to the Gordon A. Cain Center and to the Department of Mathematics at Louisiana State University for their support in the development of *Eureka Math*.

For a free *Eureka Math* Teacher Resource Pack, Parent Tip Sheets, and more please visit www.Eureka.tools

Name _____   Date _____

Read the word problem.
Draw a tape diagram or double tape diagram and label.
Write a number sentence and a statement that matches the story.

R [8]
N [8 | ?]
  12
12 − 8 = [4]

1. Peter has 3 goats living on his farm. Julio has 9 goats living on his farm. How many more goats does Julio have than Peter?

2. Willie picked 16 apples in the orchard. Emi picked 10 apples in the orchard. How many more apples did Willie pick than Emi?

3. Lee collected 13 eggs from the hens in the barn. Ben collected 18 eggs from the hens in the barn. How many fewer eggs did Lee collect than Ben?

---

4. Shanika did 14 cartwheels during recess. Kim did 20 cartwheels. How many more cartwheels did Kim do than Shanika?

EUREKA
MATH™

Name _____   Date _____

<u>R</u>ead the word problem.
<u>D</u>raw a tape diagram or double tape diagram and label.
<u>W</u>rite a number sentence and a statement that matches the story.

R [8   ]
N [8   (?)]
        └─12─┘
$12 - 8 = \boxed{4}$

1. Fran donated 11 of her old books to the library. Darnel donated 8 of his old books to the library. How many more books did Fran donate than Darnel?

_____

2. During recess, 7 students were reading books. There were 17 students playing on the playground. How many fewer students were reading books than playing on the playground?

3. Maria is 18 years old. Her brother Nikil is 12 years old. How much older is Maria than her brother Nikil?

4. It rained 15 days in the month of March. It rained 19 days in April. How many more days did it rain in April than in March?

EUREKA
MATH™

Name _____   Date _____

<u>R</u>ead the word problem.
<u>D</u>raw a tape diagram or double tape diagram and label.
<u>W</u>rite a number sentence and a statement that matches the story.

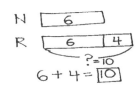

1. Nikil baked 5 pies for the contest. Peter baked 3 more pies than Nikil.
    How many pies did Peter bake for the contest?

_____

2. Emi planted 12 flowers. Rose planted 3 fewer flowers than Emi.
    How many flowers did Rose plant?

_____

3. Ben scored 15 goals in the soccer game. Anton scored 11 goals.
    How many more goals did Ben score than Anton?

_____

©2015 Great Minds. eureka-math.org
G1-M5M6-SE-B4-1.3.1-01.2016

4. Kim grew 12 roses in a garden. Fran grew 6 fewer roses than Kim.
    How many roses did Fran grow in the garden?

_____

5. Maria has 4 more fish in her tank than Shanika. Shanika has 16 fish.
    How many fish does Maria have in her tank?

_____

6. Lee has 11 board games. Lee has 5 more board games than Darnel.
    How many board games does Darnel have?

EUREKA
MATH™

Name _____   Date _____

Read the word problem.
Draw a tape diagram or double tape diagram and label.
Write a number sentence and a statement that matches the story.

N [6]
R [6 | 4]
      ?=10
6 + 4 = [10]

1. Kim went to 15 baseball games this summer. Julio went to 10 baseball games.
    How many more games did Kim go to than Julio?

_____

2. Kiana picked 14 strawberries at the farm. Tamra picked 5 fewer strawberries than
    Kiana. How many strawberries did Tamra pick?

_____

3. Willie saw 7 reptiles at the zoo. Emi saw 4 more reptiles at the zoo than Willie.
    How many reptiles did Emi see at the zoo?

4. Peter jumped into the swimming pool 6 times more than Darnel. Darnel jumped in 9 times. How many times did Peter jump into the swimming pool?

_____

5. Rose found 16 seashells on the beach. Lee found 6 fewer seashells than Rose. How many seashells did Lee find on the beach?

_____

6. Shanika got 12 cards in the mail. Nikil got 5 more cards than Shanika. How many cards did Nikil get?

Name _____ Date _____

Write the tens and ones. Complete the statement.

1.	tens	ones

43 = _____ tens _____ ones

2.	tens	ones

_____ = _____ tens _____ ones

3.	tens	ones

There are _____ cubes.

4.	tens	ones

There are _____ cubes.

5.	tens	ones

There are _____ cubes.

6.	tens	ones

There are _____ cubes.

7.	tens	ones

There are _____ peanuts.

8.	tens	ones

There are _____ juice boxes.

Lesson 3: Use the place value chart to record and name tens land ones within a two-digit number up to 100.

9

©2015 Great Minds. eureka-math.org
G1-M5M6-SE-B4-1.3.1-01.2016

9. Write the number as tens and ones in the place value chart, or use the place value chart to write the number.

a. 40

tens	ones

b. 46

tens	ones

c. _____

tens	ones
5	9

d. _____

tens	ones
9	5

e. 75

tens	ones

f. 70

tens	ones

g. 60

tens	ones

h. _____

tens	ones
8	0

i. _____

tens	ones
5	5

j. _____

tens	ones
10	0

Lesson 3:   Use the place value chart to record and name tens and ones within a
                two-digit number up to 100.

EUREKA
MATH

Name _____ Date _____

Write the tens and ones. Complete the statement.

1. 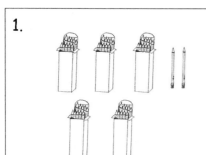	tens	ones

52 = _____ tens _____ ones

2. 	tens	ones

_____ = _____ tens _____ ones

3. 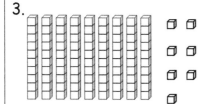 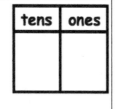	tens	ones

There are _____ cubes.

4. 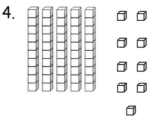 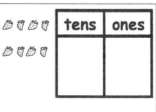	tens	ones

There are _____ cubes.

5. 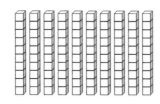	tens	ones

There are _____ cubes.

6. 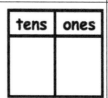 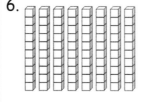	tens	ones

There are _____ cubes.

7. 	tens	ones

There are _____ carrots.

8. 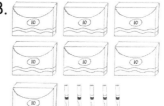	tens	ones

There are _____ markers.

EUREKA
MATH™

Lesson 3: Use the place value chart to record and name tens and ones within a two-digit number up to 100.

11

©2015 Great Minds. eureka-math.org
G1-M5M6-SE-B4-1.3.1-01.2016

9. Write the number as tens and ones in the place value chart, or use the place value chart to write the number.

a. 70

tens	ones

b. 76

tens	ones

c. _____

tens	ones
4	9

d. _____

tens	ones
9	4

e. 65

tens	ones

f. 60

tens	ones

g. 90

tens	ones

h. _____

tens	ones
10	0

i. _____

tens	ones
8	3

j. _____

tens	ones
8	0

Lesson 3:   Use the place value chart to record and name tens and ones within a two-digit number up to 100.

EUREKA MATH™

ones	tens

ones	tens

place value chart

Lesson 3:   Use the place value chart to record and name tens and ones within a
             two-digit number up to 100.

13

©2015 Great Minds. eureka-math.org
G1-M5M6-SE-B4-1.3.1-01.2016

This page intentionally left blank

Name _____   Date _____

Count the objects, and fill in the number bond or place value chart. Complete the sentences to add the tens and ones.

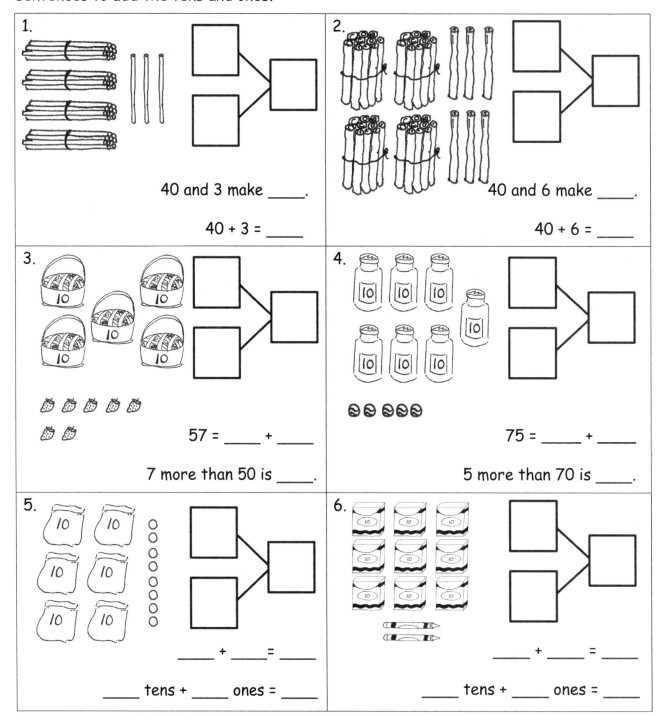

1.

40 and 3 make _____.

40 + 3 = _____

2.

40 and 6 make _____.

40 + 6 = _____

3.

57 = _____ + _____

7 more than 50 is _____.

4.

75 = _____ + _____

5 more than 70 is _____.

5.

_____ + _____ = _____

_____ tens + _____ ones = _____

6.

_____ + _____ = _____

_____ tens + _____ ones = _____

Lesson 4:   Write and interpret two-digit numbers to 100 as addition sentences
that combine tens and ones.

15

©2015 Great Minds. eureka-math.org
G1-M5M6-SE-B4-1.3.1-01.2016

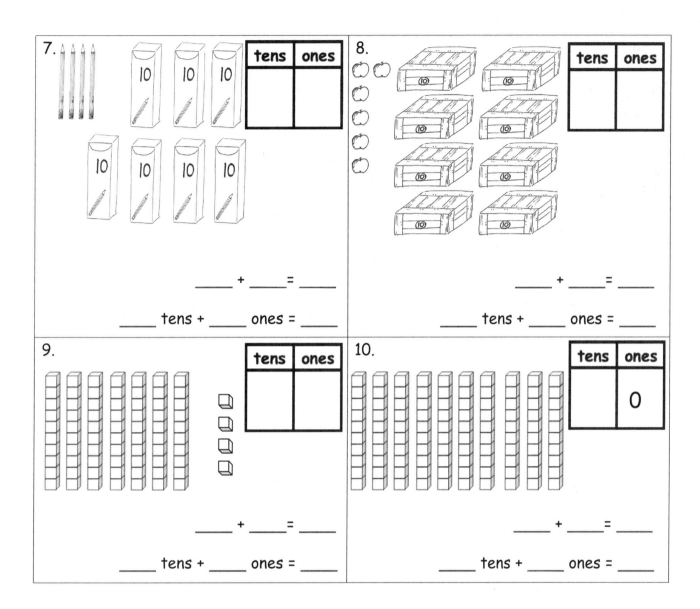

7.

tens	ones

_____ + _____ = _____

_____ tens + _____ ones = _____

8.

tens	ones

_____ + _____ = _____

_____ tens + _____ ones = _____

9.

tens	ones

_____ + _____ = _____

_____ tens + _____ ones = _____

10.

tens	ones
	0

_____ + _____ = _____

_____ tens + _____ ones = _____

11. Complete the sentences to add the tens and ones.

a. 50 + 6 = _____

b. _____ + 9 = 89

c. 5 tens + _____ ones = 56

d. 9 ones + 8 tens = _____

Lesson 4:   Write and interpret two-digit numbers to 100 as addition sentences
              that combine tens and ones.

EUREKA MATH

Name _____   Date _____

Count the objects, and fill in the number bond or place value chart. Complete the sentences to add the tens and ones.

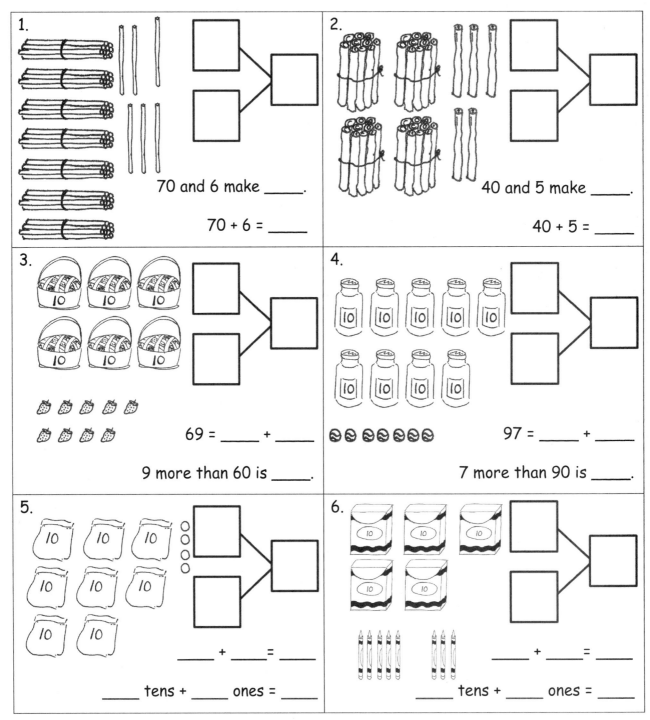

1.

70 and 6 make _____.

70 + 6 = _____

2.

40 and 5 make _____.

40 + 5 = _____

3.

69 = _____ + _____

9 more than 60 is _____.

4.

97 = _____ + _____

7 more than 90 is _____.

5.

_____ + _____ = _____

_____ tens + _____ ones = _____

6.

_____ + _____ = _____

_____ tens + _____ ones = _____

EUREKA MATH™   **Lesson 4:**   Write and interpret two-digit numbers to 100 as addition sentences that combine tens and ones.   17

©2015 Great Minds. eureka-math.org
G1-M5M6-SE-B4-1.3.1-01.2016

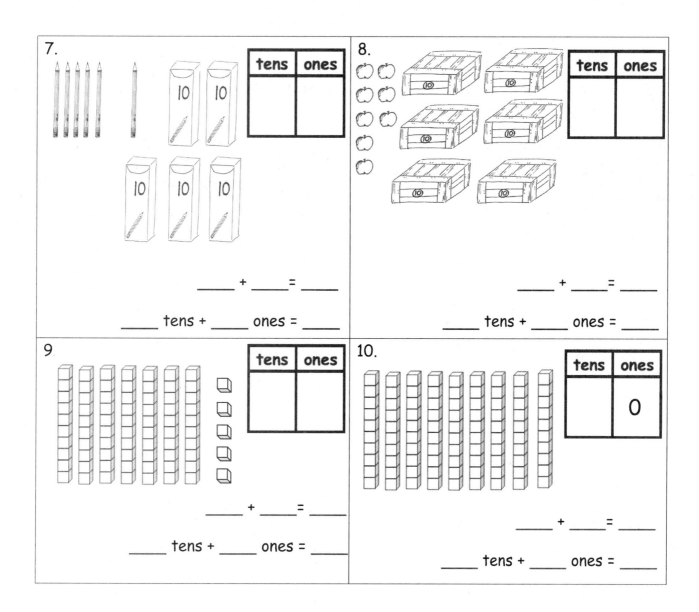

7.

tens	ones

_____ + _____ = _____

_____ tens + _____ ones = _____

8.

tens	ones

_____ + _____ = _____

_____ tens + _____ ones = _____

9.

tens	ones

_____ + _____ = _____

_____ tens + _____ ones = _____

10.

tens	ones
	0

_____ + _____ = _____

_____ tens + _____ ones = _____

11. Complete the sentences to add the tens and ones.

a. 80 + 6 = _____

b. _____ + 7 = 57

c. 9 tens + _____ ones = 95

d. 4 ones + 8 tens = _____

Lesson 4: Write and interpret two-digit numbers to 100 as addition sentences that combine tens and ones.

EUREKA MATH

Name _____   Date _____

1. Solve. You may draw or cross off (x) to show your work.

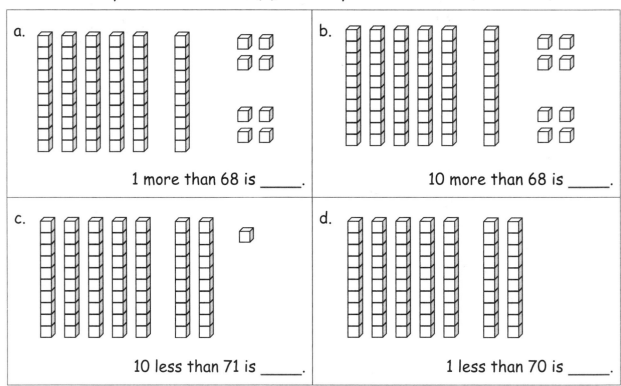

a.

1 more than 68 is _____.

b.

10 more than 68 is _____.

c.

10 less than 71 is _____.

d.

1 less than 70 is _____.

2. Find the mystery numbers. Use the arrow way to explain how you know.

a. 10 more than 59 is _____.

tens	ones
5	9

+1 ten →

tens	ones

b. 1 less than 59 is _____.

tens	ones

tens	ones

c. 1 more than 59 is _____.

tens	ones

tens	ones

d. 10 less than 59 is _____.

tens	ones

tens	ones

3. Write the number that is **1 more**.

    a. 10, 11⎯   †

    b. 70, 71⎯

    c. 76, 77⎯

    d. 79, 80⎯

    e. 99, 100⎯

4. Write the number that is **10 more**.

    a. 10, ~~20~~ 20

    b. 60, ~~120~~ 80

    c. 61, ⎯ 81

    d. 78, ⎯ 98

    e. 90, ⎯ 100

5. Write the number that is **1 less**.

    a. 12, 11⎯   ⎯

    b. 52, 51⎯

    c. 51, 50⎯

    d. 80, 79⎯

    e. 100, 99⎯

6. Write the number that is **10 less**.

    a. 20, ⎯

    b. 60, ⎯

    c. 74, ⎯

    d. 81, ⎯

    e. 100, ⎯

7. Fill in the missing numbers in each sequence.

    a. 40, 41, 42, ⎯⎯

    b. 89, 88, 87, ⎯⎯

    c. 72, 71, ⎯⎯, 69

    d. 63, ⎯⎯, 65, 66

    e. 40, 50, 60, ⎯⎯

    f. 80, 70, 60, ⎯⎯

    g. 55, 65, ⎯⎯, 85

    h. 99, 89, ⎯⎯, 69

    i. ⎯⎯, 99, 98, 97

    j. ⎯⎯, 77, ⎯⎯, 57

Lesson 5:   Identify 10 more, 10 less, 1 more, and 1 less than a two-digit number within 100.

EUREKA
MATH™

Name _____   Date _____

1. Solve. You may draw or cross off (x) to show your work.

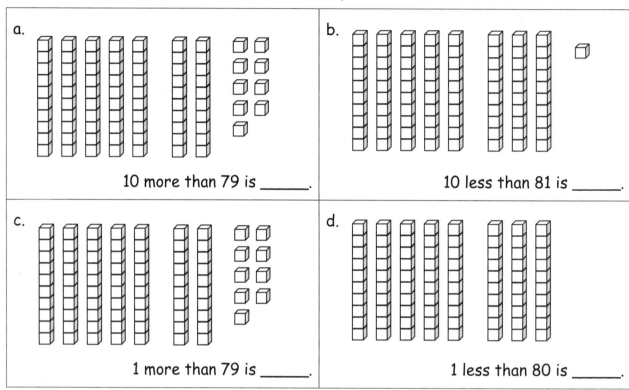

a.

10 more than 79 is _____.

b.

10 less than 81 is _____.

c.

1 more than 79 is _____.

d.

1 less than 80 is _____.

2. Find the mystery numbers. You may make a drawing to help solve, if needed.

a. 10 more than 75 is _____.

tens	ones
7	5

+ 10 →

tens	ones

b. 1 more than 75 is _____.

tens	ones

→

tens	ones

c. 10 less than 88 is _____.

tens	ones

tens	ones

d. 1 less than 88 is _____.

tens	ones

tens	ones

EUREKA
MATH™

Lesson 5:   Identify 10 more, 10 less, 1 more, and 1 less than a two-digit number within 100.

21

3. Write the number that is **1 more**.

    a. 40, _____

    b. 50, _____

    c. 65, _____

    d. 69, _____

    e. 99, _____

4. Write the number that is **10 more**.

    a. 60, _____

    b. 70, _____

    c. 77, _____

    d. 89, _____

    e. 90, _____

5. Write the number that is **1 less**.

    a. 53, _____

    b. 73, _____

    c. 71, _____

    d. 80, _____

    e. 100, _____

6. Write the number that is **10 less**.

    a. 50, _____

    b. 60, _____

    c. 84, _____

    d. 91, _____

    e. 100, _____

7. Fill in the missing numbers in each sequence.

    a. 50, 51, 52, _____

    b. 79, 78, 77, _____

    c. 62, 61, _____, 59

    d. 83, _____, 85, 86

    e. 60, 70, 80, _____

    f. 100, 90, 80, _____

    g. 57, 67, _____, 87

    h. 89, 79, _____, 59

    i. _____, 99, 98, 97

    j. _____, 84, _____, 64

EUREKA
MATH™

Name _____   Date _____

1. Use the symbols to compare the numbers. Fill in the blank with <, >, or = to make the statement true.

   85   75      4 tens 3 ones   4 tens 6 ones

85 (>) 75   43 (<) 46
85 is greater than 75.   43 is less than 46.

| a. 35 ◯ 42 | b. 78 ◯ 80 |

| c. 100 ◯ 99 | d. 93 ◯ 8 tens 3 ones |

| e. 9 tens 8 ones ◯ 10 tens | f. 6 tens 2 ones ◯ 2 tens 6 ones |

| g. 72 ◯ 2 ones 7 tens | h. 5 tens 4 ones ◯ 4 tens 14 ones |

Lesson 6: Use the symbols >, =, and < to compare quantities and numerals to 100.

23

EUREKA MATH™

2. Circle the correct words to make the sentence true. Use >, <, or = and numbers to write a true statement.

a.		
29	is greater than is less than is equal to	2 tens 9 ones

_____ ◯ _____

b.		
7 tens 9 ones	is greater than is less than is equal to	80

_____ ◯ _____

c.		
10 tens 0 ones	is greater than is less than is equal to	0 tens 10 ones

_____ ◯ _____

d.		
6 tens 1 one	is greater than is less than is equal to	5 tens 16 ones

_____ ◯ _____

3. Use <, =, or > to compare the pairs of numbers.

a.   3 tens 9 ones ◯ 5 tens 9 ones

b.   30 ◯ 13

c.   100 ◯ 10 tens

d.   6 tens 4 ones ◯ 4 ones 6 tens

e.   7 tens 9 ones ◯ 79

f.   1 ten 5 ones ◯ 5 ones 1 ten

g.   72 ◯ 6 tens 12 ones

h.   88 ◯ 8 tens 18 ones

Lesson 6:   Use the symbols >, =, and < to compare quantities and numerals to 100.

EUREKA MATH

Name _____   Date _____

1. Use the symbols to compare the numbers. Fill in the blank with <, >, or = to make the statement true.

62   57

5 tens 6 ones   5 tens 9 ones

62 (>) 57   56 (<) 59

62 is greater than 57.   56 is less than 59.

a.		b.	
43 ( ) 35		60 ( ) 86	

c.		d.	
10 tens ( ) 99		5 tens 4 ones ( ) 54	

e.		f.	
7 tens 9 ones ( ) 9 tens 7 ones		1 ten 3 ones ( ) 31	

g.		h.	
3 tens 0 ones ( ) 2 tens 10 ones		3 tens 5 ones ( ) 2 tens 17 ones	

2. Fill in the correct words from the box to make the sentence true. Use >, <, or = and numbers to write a true statement.

is greater than	is less than	is equal to

a. 42 _____ 1 ten 2 ones

____ ◯ ____

b. 6 tens 7 ones _____ 5 tens 17 ones

____ ◯ ____

c. 37 _____ 73

____ ◯ ____

d. 2 tens 14 ones _____ 4 ones 2 tens

____ ◯ ____

e. 9 ones 5 tens _____ 9 tens 5 ones

____ ◯ ____

Lesson 6: Use the symbols >, =, and < to compare quantities and numerals to 100.

EUREKA MATH™

©2015 Great Minds. eureka-math.org
G1-M5M6-SE-B4-1.3.1-01.2016

Name _____   Date _____

1. Fill in the missing numbers in the chart up to 120.

a.	b.	c.	d.	e.
71	81	91		111
	82		102	
73	83	93		113
	84	94	104	114
76	86	96	106	116
77	87	97		117
79	89	99	109	119
80		100	110	

Lesson 7:   Count and write numbers to 120. Use Hide Zero cards to relate
numbers 0 to 20 to 100 to 120.

27

©2015 Great Minds. eureka-math.org
G1-M5M6-SE-B4-1.3.1-01.2016

2. Write the numbers to continue the counting sequence to 120.

96, 97, _____, _____, _____, _____, _____,

_____, _____, _____, _____, _____, _____,

_____, _____, _____, _____, _____, _____,

_____, _____, _____, _____, _____, _____

3. Circle the sequence that is incorrect. Rewrite it correctly on the line.

a.

| 107, 108, 109, 110, 120 |

b.

| 99, 100, 101, 102, 103 |

_____

4. Fill in the missing numbers in the sequence.

a.

| 115, 116, _____, _____, _____ |

b.

| _____, _____, 118, _____, 120 |

c.

| 100, 101, _____, _____, 104 |

d.

| 97, 98, _____, _____, _____, _____ |

EUREKA
MATH™

Name _____ Date _____

1. Fill in the missing numbers in the chart up to 120.

a.	b.	c.	d.	e.
71		91		111
	82		102	
		93		
74				114
	85		105	
		96		116
	87			
			108	
79		99		119
80	90		110	

Lesson 7: Count and write numbers to 120. Use Hide Zero cards to relate numbers 0 to 20 to 100 to 120.

29

©2015 Great Minds. eureka-math.org
G1-M5M6-SE-B4-1.3.1-01.2016

2. Write the numbers to continue the counting sequence to 120.

99, _____, 101, _____, _____, _____, _____, _____, _____,

_____, _____, _____, _____, _____, _____, _____,

_____, _____, _____, _____, _____, _____

3. Circle the sequence that is incorrect. Rewrite it correctly on the line.

a.   b.

| 116, 117, 118, 119, 120 | | 96, 97, 98, 99, 100, 110 |

_____

4. Fill in the missing numbers in the sequence.

a.   b.

| 113, 114, _____, _____, _____ | | _____, _____, _____, 120 |

c.   d.

| 102, _____, _____, _____ | | 88, 89, _____, _____, _____, _____ |

EUREKA
MATH

Name _____   Date _____

1. Write the number as tens and ones in the place value chart, or use the place value chart to write the number.

a. 74

tens	ones

b. 78

tens	ones

c. _____

tens	ones
9	1

d. _____

tens	ones
10	9

e. 116

tens	ones

f. 103

tens	ones

g. _____

tens	ones
11	2

h. _____

tens	ones
12	0

i. _____

tens	ones
10	5

j. 102

tens	ones

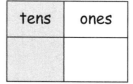

Lesson 8:   Count to 120 in unit form using only tens and ones. Represent
             numbers to 120 as tens and ones on the place value chart.

31

©2015 Great Minds. eureka-math.org
G1-M5M6-SE-B4-1.3.1-01.2016

2. Match.

a.

tens	ones
9	7

●   ● 10 tens 5 ones

b.

tens	ones
10	7

●   ● 10 tens 7 ones

c.

tens	ones
11	0

●   ● 9 tens 7 ones

d.

tens	ones
10	5

●   ● 12 tens 0 ones

e.

tens	ones
10	1

●   ● 110

f.

tens	ones
12	0

●   ● 11 tens 8 ones

g.

tens	ones
11	8

●   ● 101

Lesson 8: Count to 120 in unit form using only tens and ones. Represent numbers to 120 as tens and ones on the place value chart.

EUREKA MATH

Name _____ Date _____

1. Write the number as tens and ones in the place value chart, or use the place value chart to write the number.

a. 81

tens	ones

b. 98

tens	ones

c. _____

tens	ones
11	7

d. _____

tens	ones
10	8

e. 104

tens	ones

f. 111

tens	ones

2. Write the number.

a. 9 tens 2 ones is the number _____.	b. 8 tens 4 ones is the number _____.
c. 11 tens 3 ones is the number _____.	d. 10 tens 9 ones is the number _____.
e. 10 tens 1 ones is the number _____.	f. 11 tens 6 ones is the number _____.

Lesson 8: Count to 120 in unit form using only tens and ones. Represent numbers to 120 as tens and ones on the place value chart.

33

3. Match.

a.

tens	ones
10	2

●   ●   11 tens 4 ones

b.

tens	ones
9	5

●   ●   9 tens 5 ones

c.

tens	ones
11	4

●   ●   11 tens 8 ones

d.

tens	ones
11	0

●   ●   11 tens 0 ones

e.

tens	ones
10	8

●   ●   102

f.

tens	ones
10	0

●   ●   10 tens 0 ones

g.

tens	ones
11	8

●   ●   108

Lesson 8:   Count to 120 in unit form using only tens and ones. Represent numbers to 120 as tens and ones on the place value chart.

©2015 Great Minds. eureka-math.org
G1-M5M6-SE-B4-1.3.1-01.2016

EUREKA
MATH

Name _____   Date _____

Count the objects. Fill in the place value chart, and write the number on the line.

1.

tens	ones

_____

2.

tens	ones

_____

3.

tens	ones

_____

4.

tens	ones

_____

5.

tens	ones

_____

6.

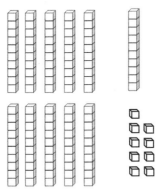

tens	ones

_____

7.

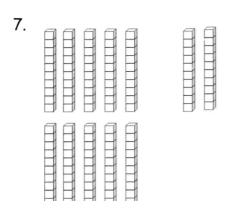

tens	ones

_____

Use quick tens and ones to represent the following numbers. Write the number on the line.

8. _____

tens	ones
10	9

9. _____

tens	ones
12	0

Lesson 9:   Represent up to 120 objects with a written numeral.

EUREKA
MATH™

Name _____   Date _____

Count the objects. Fill in the place value chart, and write the number on the line.

1.

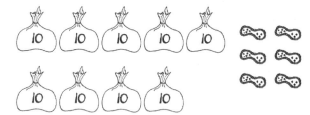

tens	ones

_____

2.

tens	ones

_____

3.

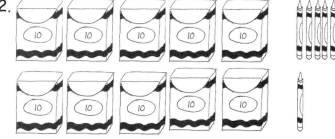

tens	ones

_____

4.

tens	ones

_____

5.

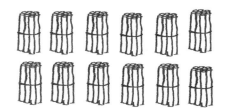

tens	ones

_____

EUREKA MATH

Lesson 9:   Represent up to 120 objects with a written numeral.

37

6.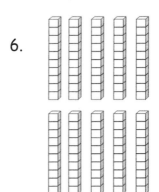

tens	ones

_____

7.

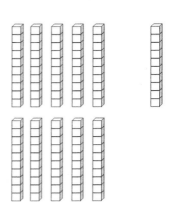

tens	ones

_____

Use quick tens and ones to represent the following numbers.

Write the number on the line.

8. _____

tens	ones
11	0

9. _____

tens	ones
10	5

EUREKA MATH

Name _____   Date _____

Complete the number bonds and number sentences to match the picture.

1.

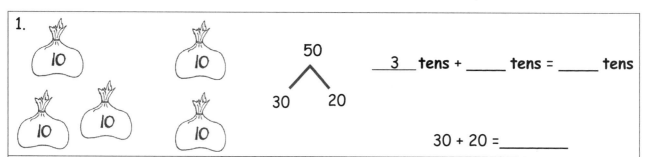

___3___ tens + _____ tens = _____ tens

30 + 20 = _____

2.

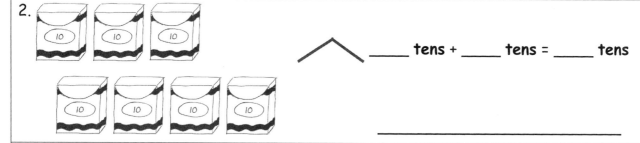

_____ tens + _____ tens = _____ tens

_____

3.

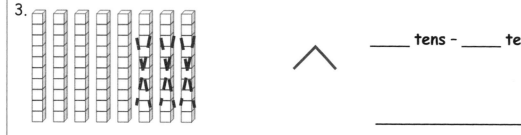

_____ tens – _____ tens = _____ tens

_____

4.

_____ tens + _____ tens = _____ tens

_____

5.

_____ tens – _____ tens = _____ tens

_____

Lesson 10:   Add and subtract multiples of 10 from multiples of 10 to 100, including dimes.

39

Count the dimes to add or subtract. Write a number sentence to match the value of the dimes.

6.   +   40 + 20 = _____

7.   _____

8.   +

                            _____

9.

             _____

10.

              _____

11. Fill in the missing numbers.

   a. 40 + 40 = _____   b. 50 − 30 = _____   c. 10 + _____ = 70

   d. 60 − _____ = 0   e. 90 − _____ = 10   f. 70 + _____ = 90

   g. 50 + 40 = _____   h. 100 − 30 = _____   i. 100 − _____ = 70

Lesson 10:   Add and subtract multiples of 10 from multiples of 10 to 100, including dimes.   **EUREKA MATH**

Name _____   Date _____

1. Complete the number bond or number sentence, and draw a line to the matching picture.

a.

90
30 [ ]

b.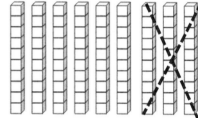

_____ – 40 = 60

c.

60
40 [ ]

d.

80 – _____ = 60

2. Count the dimes to add or subtract. Write a number sentence to match the dimes.

a.   +

$$40 + 20 = \underline{\hspace{4cm}}$$

b.

_____

c.  +

_____

d.

_____

3. Fill in the missing numbers.

a. 70 + _____ = 90

b. _____ + 30 = 80

c. 100 – _____ = 20

d. 30 + 60 = _____

e. 70 – _____ = 20

f. 20 + _____ = 60

g. _____ – 20 = 60

h. 90 – _____ = 20

i. 50 + _____ = 100

Lesson 10: Add and subtract multiples of 10 from multiples of 10 to 100, including dimes.

EUREKA MATH™

_____ tens ◯ _____ tens ◯ _____ tens

_____
number bond/number sentence set

This page intentionally left blank

Name _____ Date _____

Solve using the pictures. Complete the number sentence to match.

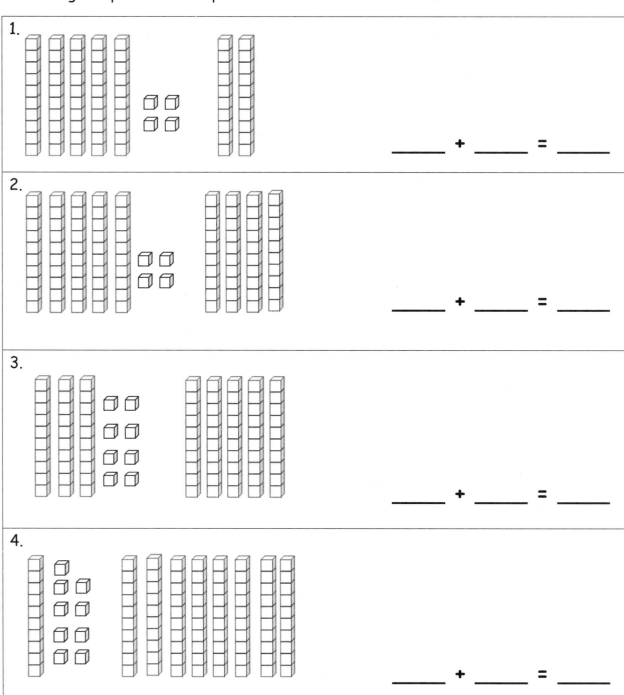

1. _____ + _____ = _____

2. _____ + _____ = _____

3. _____ + _____ = _____

4. _____ + _____ = _____

EUREKA
MATH™

Lesson 11:   Add a multiple of 10 to any two-digit number within 100.

45

5. Solve.

a. 47 + 40 = _____	b. 57 + 30 = _____
c. 35 + 30 = _____	d. 35 + 50 = _____
e. 30 + 63 = _____	f. 40 + 39 = _____

6. Solve and explain your thinking to a partner.

a. 2 + 50 = _____

b. 58 + 40 = _____

c. 48 + _____ = 98

d. 60 + _____ = 86

Lesson 11:   Add a multiple of 10 to any two-digit number within 100.

EUREKA MATH

©2015 Great Minds. eureka-math.org
G1-M5M6-SE-B4-1.3.1-01.2016

Name _____   Date _____

1. Solve using the pictures. Complete the number sentence to match.

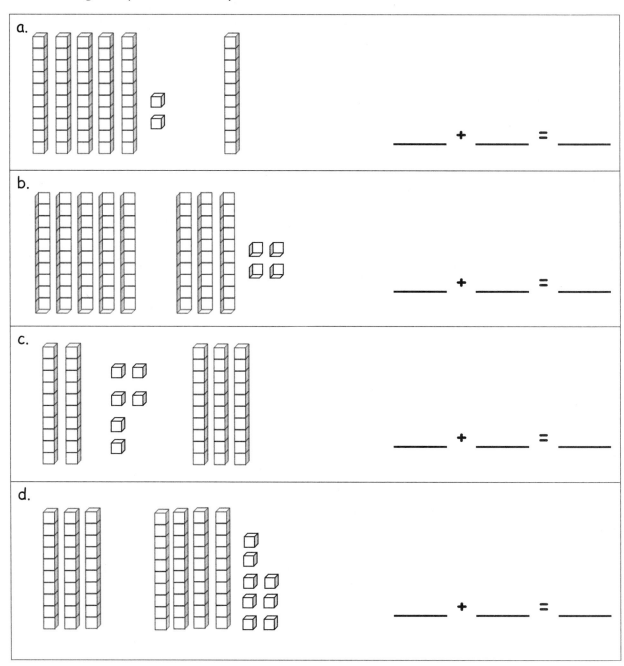

a.

_____ + _____ = _____

b.

_____ + _____ = _____

c.

_____ + _____ = _____

d.

_____ + _____ = _____

EUREKA
MATH™

Lesson 11:   Add a multiple of 10 to any two-digit number within 100.

47

©2015 Great Minds. eureka-math.org
G1-M5M6-SE-B4-1.3.1-01.2016

$$64 + 30 = 94$$
$$4 \quad 60$$
$$60 + 30 = 90$$
$$90 + 4 = 94$$

2. Use number bonds to solve.

a. 38 + 40 = _____	b. 54 + 30 = _____
c. 46 + 40 = _____	d. 30 + 57 = _____
e. 20 + 68 = _____	f. 25 + 70 = _____

3. Solve. You may use number bonds to help you.

a. 72 + 20 = _____

b. 48 + 50 = _____

c. 46 + _____ = 96

d. _____ + 40 = 87

EUREKA MATH™

Name _____   Date _____

1. Solve.

a. 84 + 12 = _____	b. 71 + 26 = _____
c. 57 + 22 = _____	d. 59 + 41 = _____
e. 35 + 65 = _____	f. 26 + 54 = _____
g. 57 + 42 = _____	h. 37 + 63 = _____

Lesson 12:   Add a pair of two-digit numbers when the ones digits have a sum less
than or equal to 10.

49

©2015 Great Minds. eureka-math.org
G1-M5M6-SE-B4-1.3.1-01.2016

2. Solve.

a. 45 + 13 = _____	b. 45 + 23 = _____
c. 21 + 27 = _____	d. 27 + 23 = _____
e. 48 + 32 = _____	f. 48 + 52 = _____
g. 34 + 65 = _____	h. 46 + 43 = _____

Lesson 12:   Add a pair of two-digit numbers when the ones digits have a sum less than or equal to 10.

EUREKA
MATH™

Name _____   Date _____

1. Solve.

a. 46 + 22 = _____	b. 74 + 23 = _____
c. 54 + 25 = _____	d. 68 + 31 = _____
e. 45 + 55 = _____	f. 86 + 13 = _____
g. 37 + 52 = _____	h. 47 + 52 = _____

EUREKA MATH

Lesson 12:   Add a pair of two-digit numbers when the ones digits have a sum less than or equal to 10.

51

2. Solve using number bonds. You may choose to add the ones or tens first. Write the two number sentences to show what you did.

a. 76 + 23 = _____	b. 45 + 33 = _____
c. 31 + 67 = _____	d. 57 + 32 = _____
e. 58 + 21 = _____	f. 25 + 63 = _____
g. 44 + 55 = _____	h. 47 + 53 = _____

Lesson 12: Add a pair of two-digit numbers when the ones digits have a sum less than or equal to 10.

©2015 Great Minds. eureka-math.org
G1-M5M6-SE-B4-1.3.1-01.2016

EUREKA MATH™

Name _____   Date _____

1. Solve and show your work.

a. 79 + 12 = _____	b. 59 + 32 = _____
c. 38 + 45 = _____	d. 36 + 47 = _____
e. 48 + 45 = _____	f. 57 + 34 = _____

EUREKA
MATH™

Lesson 13:   Add a pair of two-digit numbers when the ones digits have a sum
              greater than 10 using decomposition.

©2015 Great Minds. eureka-math.org
G1-M5M6-SE-B4-1.3.1-01.2016

53

2. Solve and show your work.

a. 24 + 37 = _____	b. 48 + 45 = _____
c. 29 + 67 = _____	d. 48 + 34 = _____
e. 69 + 27 = _____	f. 78 + 17 = _____

Lesson 13:   Add a pair of two-digit numbers when the ones digits have a sum
                                greater than 10 using decomposition.

EUREKA
MATH™

Name _____ Date _____

1. Solve and show your work.

a. 15 + 26 = _____	b. 46 + 49 = _____	c. 28 + 54 = _____
d. 69 + 13 = _____	e. 69 + 23 = _____	f. 69 + 19 = _____
g. 49 + 43 = _____	h. 57 + 36 = _____	i. 68 + 23 = _____

Lesson 13: Add a pair of two-digit numbers when the ones digits have a sum greater than 10 using decomposition.

55

©2015 Great Minds. eureka-math.org
G1-M5M6-SE-B4-1.3.1-01.2016

2. Solve and show your work.

a. 34 + 47 = _____	b. 38 + 45 = _____	c. 68 + 23 = _____
d. 39 + 57 = _____	e. 38 + 44 = _____	f. 17 + 76 = _____
g. 68 + 24 = _____	h. 18 + 77 = _____	i. 14 + 67 = _____

Lesson 13: Add a pair of two-digit numbers when the ones digits have a sum greater than 10 using decomposition.

EUREKA MATH™

Name _____   Date _____

1. Solve and show your work.

a. 48 + 21 = _____	b. 48 + 22 = _____
c. 39 + 43 = _____	d. 48 + 34 = _____
e. 77 + 14 = _____	f. 67 + 27 = _____
g. 58 + 37 = _____	h. 68 + 29 = _____

Lesson 14:   Add a pair of two-digit numbers when the ones digits have a sum greater than 10 using decomposition.

57

2. Solve and show your work.

a. 39 + 31 = _____	b. 58 + 23 = _____
c. 77 + 23 = _____	d. 69 + 26 = _____
e. 68 + 25 = _____	f. 45 + 37 = _____
g. 59 + 39 = _____	h. 58 + 38 = _____

Lesson 14:   Add a pair of two-digit numbers when the ones digits have a sum
greater than 10 using decomposition.

EUREKA
MATH™

Name _____   Date _____

1. Solve and show your work.

a. 68 + 21 = _____	b. 59 + 32 = _____
c. 39 + 44 = _____	d. 58 + 36 = _____
e. 76 + 17 = _____	f. 68 + 26 = _____
g. 56 + 39 = _____	h. 58 + 29 = _____

Lesson 14:   Add a pair of two-digit numbers when the ones digits have a sum greater than 10 using decomposition.

59

2. Solve and show your work.

a. 39 + 41 = _____	b. 48 + 43 = _____
c. 87 + 13 = _____	d. 59 + 25 = _____
e. 65 + 27 = _____	f. 27 + 67 = _____
g. 49 + 39 = _____	h. 38 + 58 = _____

Lesson 14:   Add a pair of two-digit numbers when the ones digits have a sum greater than 10 using decomposition.

EUREKA
MATH™

Name _____ Date _____

1. Solve using quick tens and ones drawings. Remember to line up your tens with tens and ones with ones. Write the total below your drawing.

a. 29 + 42 = _____	b. 39 + 54 = _____
c. 41 + 38 = _____	d. 58 + 24 = _____
e. 47 + 46 = _____	f. 48 + 29 = _____

EUREKA MATH

Lesson 15: Add a pair of two-digit numbers when the ones digits have a sum greater than 10 with drawing. Record the total below.

61

©2015 Great Minds. eureka-math.org
G1-M5M6-SE-B4-1.3.1-01.2016

2. Solve using quick tens and ones. Remember to line up your tens with tens and ones with ones. Write the total below your drawing.

a. 49 + 22 = _____	b. 38 + 62 = _____
c. 59 + 23 = _____	d. 68 + 14 = _____
e. 46 + 36 = _____	f. 69 + 26 = _____

Lesson 15:   Add a pair of two-digit numbers when the ones digits have a sum greater than 10 with drawing. Record the total below.

Name _____   Date _____

1. Solve using quick tens and ones drawings. Remember to line up your tens with tens and ones with ones. Write the total below your drawing.

a. 39 + 42 = _____	b. 48 + 36 = _____
c. 31 + 48 = _____	d. 47 + 34 = _____
e. 57 + 39 = _____	f. 58 + 27 = _____

2. Solve using quick tens and ones. Remember to line up your tens with tens and ones with ones. Write the total below your drawing.

a. 59 + 25 = _____	b. 48 + 42 = _____
c. 39 + 53 = _____	d. 78 + 14 = _____
e. 57 + 25 = _____	f. 69 + 27 = _____

Lesson 15:   Add a pair of two-digit numbers when the ones digits have a sum greater than 10 with drawing. Record the total below.

EUREKA MATH

Name _____ Date _____

1. Solve using quick tens and ones drawings. Remember to line up your drawings and rewrite the number sentence vertically.

a. 29 + 43 = _____ 29 + 43 ———— 72 72	b. 34 + 49 = _____
c. 45 + 39 = _____	d. 54 + 25 = _____
e. 47 + 36 = _____	f. 54 + 46 = _____

Lesson 16:   Add a pair of two-digit numbers when the ones digits have a sum
greater than 10 with drawing. Record the new ten below.

65

©2015 Great Minds. eureka-math.org
G1-M5M6-SE-B4-1.3.1-01.2016

2. Solve using quick tens and ones. Remember to line up your drawings and rewrite the number sentence vertically.

a. 39 + 24 = _____	b. 58 + 36 = _____
c. 55 + 37 = _____	d. 59 + 36 = _____
e. 37 + 58 = _____	f. 68 + 29 = _____

Lesson 16:   Add a pair of two-digit numbers when the ones digits have a sum greater than 10 with drawing. Record the new ten below.

EUREKA MATH™

©2015 Great Minds. eureka-math.org
G1-M5M6-SE-B4-1.3.1-01.2016

Name _____   Date _____

1. Solve using quick tens and ones drawings. Remember to line up
    your drawings and rewrite the number sentence vertically.

a. 39 + 45 = _____	b. 64 + 28 = _____
c. 47 + 38 = _____	d. 53 + 27 = _____
e. 38 + 48 = _____	f. 53 + 45 = _____

Lesson 16:   Add a pair of two-digit numbers when the ones digits have a sum
               greater than 10 with drawing. Record the new ten below.

2. Solve using quick tens and ones. Remember to line up your drawings and rewrite the number sentence vertically.

a. 79 + 14 = _____	b. 28 + 47 = _____
c. 58 + 33 = _____	d. 19 + 66 = _____
e. 39 + 59 = _____	f. 49 + 48 = _____

Lesson 16: Add a pair of two-digit numbers when the ones digits have a sum greater than 10 with drawing. Record the new ten below.

EUREKA
MATH™

Name _____   Date _____

1. Solve using quick tens and ones drawings. Remember to line up your tens and ones and rewrite the number sentence vertically.

a. 39 + 52 = _____	b. 48 + 42 = _____
c. 47 + 42 = _____	d. 47 + 47 = _____
e. 68 + 17 = _____	f. 68 + 29 = _____

Lesson 17:   Add a pair of two-digit numbers when the ones digits have a sum greater than 10 with drawing. Record the new ten below.

69

©2015 Great Minds. eureka-math.org
G1-M5M6-SE-B4-1.3.1-01.2016

2. Solve using quick tens and ones drawings. Remember to line up your tens and ones and rewrite the number sentence vertically.

a. 39 + 32 = _____	b. 48 + 31 = _____
c. 43 + 49 = _____	d. 57 + 38 = _____
e. 61 + 39 = _____	f. 68 + 25 = _____

Lesson 17: Add a pair of two-digit numbers when the ones digits have a sum greater than 10 with drawing. Record the new ten below.

EUREKA MATH

©2015 Great Minds. eureka-math.org
G1-M5M6-SE-B4-1.3.1-01.2016

Name _____   Date _____

1. Solve using quick tens and ones drawings. Remember to line up your tens and ones and rewrite the number sentence vertically.

a. 49 + 33 = _____	b. 68 + 32 = _____
c. 36 + 43 = _____	d. 27 + 67 = _____
e. 78 + 17 = _____	f. 69 + 28 = _____

Lesson 17:   Add a pair of two-digit numbers when the ones digits have a sum greater than 10 with drawing. Record the new ten below.

71

2. Solve using quick tens and ones drawings. Remember to line up your tens and ones and rewrite the number sentence vertically.

a. 29 + 52 = _____	b. 58 + 31 = _____
c. 73 + 26 = _____	d. 67 + 28 = _____
e. 41 + 59 = _____	f. 48 + 45 = _____

Lesson 17:   Add a pair of two-digit numbers when the ones digits have a sum greater than 10 with drawing. Record the new ten below.

EUREKA
MATH™

Name _____   Date _____

Use any method you prefer to solve the problems below.

1. 74 + 21 = _____	2. 79 + 21 = _____
3. 46 + 34 = _____	4. 58 + 34 = _____
5. 35 + 14 = _____	6. 35 + 18 = _____

Lesson 18:   Add a pair of two-digit numbers with varied sums in the ones, and
compare the results of different recording methods.

73

©2015 Great Minds. eureka-math.org
G1-M5M6-SE-B4-1.3.1-01.2016

Name _____ Date _____

Use any method you prefer to solve the problems below.

1. 61 + 15 = _____	2. 16 + 51 = _____
3. 37 + 45 = _____	4. 27 + 46 = _____
5. 58 + 27 = _____	6. 38 + 48 = _____

Lesson 18:   Add a pair of two-digit numbers with varied sums in the ones, and compare the results of different recording methods.

EUREKA MATH™

Name _____   Date _____

Use the strategy you prefer to solve the problems below.

1. 43 + 21 = _____	2. 43 + 41 = _____
3. 62 + 38 = _____	4. 52 + 48 = _____
5. 75 + 14 = _____	6. 75 + 16 = _____

Lesson 19:   Solve and share strategies for adding two-digit numbers with varied sums.

©2015 Great Minds. eureka-math.org
G1-M5M6-SE-B4-1.3.1-01.2016

75

Use the strategy you prefer to solve the problems below.

7. 29 + 54 = _____	8. 27 + 54 = _____
9. 38 + 23 = _____	10. 58 + 36 = _____
11. 49 + 19 = _____	12. 28 + 69 = _____

Lesson 19:   Solve and share strategies for adding two-digit numbers with varied sums.

EUREKA
MATH™

Name _____   Date _____

Use the strategy you prefer to solve the problems below.

1. 53 + 22 = _____	2. 23 + 52 = _____
3. 76 + 14 = _____	4. 76 + 16 = _____
5. 55 + 35 = _____	6. 54 + 46 = _____

EUREKA
MATH™

Lesson 19:   Solve and share strategies for adding two-digit numbers with varied
sums.

77

©2015 Great Minds. eureka-math.org
G1-M5M6-SE-B4-1.3.1-01.2016

Use the strategy you prefer to solve the problems below.

7. 49 + 25 = _____	8. 49 + 45 = _____
9. 37 + 37 = _____	10. 37 + 57 = _____
11. 24 + 48 = _____	12. 26 + 68 = _____

Lesson 19: Solve and share strategies for adding two-digit numbers with varied sums.

©2015 Great Minds. eureka-math.org
G1-M5M6-SE-B4-1.3.1-01.2016

EUREKA
MATH™

Name _____   Date _____

1. Use the word bank to label the coin. The front and back of the coin is shown.

| penny |
| nickel |
| dime |

a. _____   b. _____   c. _____

2. Draw more pennies to show the value of each coin.

a.

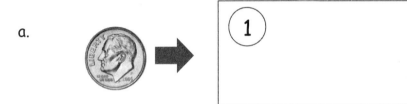
→ ①

b.

→ ①

3. Kim has 5 cents in her hand. Cross off (x) the hand that cannot be Kim's.

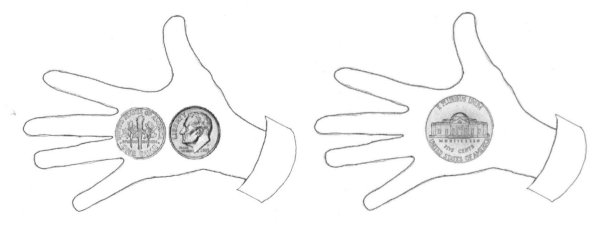

EUREKA
MATH™

Lesson 20:   Identify pennies, nickels, and dimes by their image, name, or value.
Decompose the values of nickels and dimes using pennies and nickels.

79

©2015 Great Minds. eureka-math.org
G1-M5M6-SE-B4-1.3.1-01.2016

4. Anton has 10 cents in his pocket. One of his coins is a nickel. Draw coins to show two different ways he could have ten cents with the coins he has in his pocket.

5. Emi says she has more money than Kiana. Is she correct? Why or why not?

Emi's Money

Kiana's Money

Emi is correct/not correct because _____

_____

EUREKA
MATH™

Name _____   Date _____

1. Match.

| penny |

| nickel |

| dime |

2. Cross off some pennies so the remaining pennies show the value of the coin to their left.

a.

b.

3. Maria has 5 cents in her pocket. Draw coins to show two different ways she could have 5 cents.

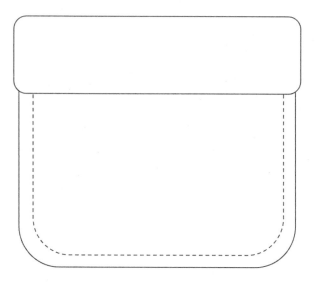   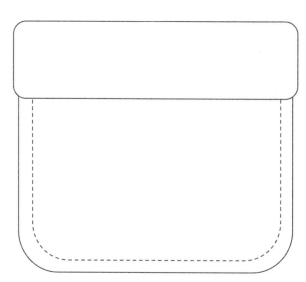

4. Solve. Draw a line to match the number sentence with the coin (or coins) that give the answer.

   a. 10 cents + 10 cents = _____ cents   ●   ●

   b. 10 cents - 5 cents = _____ cents   ●   ●

   c. 20 cents – 10 cents = _____ cents   ●   ●

   d. 9 cents – 8 cents = _____ cents   ●   ●

Lesson 20:   Identify pennies, nickels, and dimes by their image, name, or value.
                         Decompose the values of nickels and dimes using pennies and nickels.

EUREKA
MATH™

Name _____   Date _____

1. Use different coin combinations to make 25 cents.

a.

_____ pennies

b.

_____ dimes

_____ pennies

c.

_____ dimes

_____ nickels

d.

_____ nickels

_____ pennies

e.

_____ nickels

f.

_____ quarter

Lesson 21:   Identify quarters by their image, name, or value. Decompose the value
of a quarter using pennies, nickels, and dimes.

83

©2015 Great Minds. eureka-math.org
G1-M5M6-SE-B4-1.3.1-01.2016

2. Use the word bank to label the coins.

| pennies   nickels   dimes   quarters |

a. _____   b. _____   c. _____   d. _____

3. Draw different coins to show the value of the coin shown.

4. Match the coin combinations to the coin with the same value.

a.    •   •

b.    •   •

c.    •   •

Lesson 21:   Identify quarters by their image, name, or value. Decompose the value
of a quarter using pennies, nickels, and dimes.

EUREKA MATH

Name _____   Date _____

1. Use the word bank to label the coins.

| dimes   nickels   pennies   quarters |

a. _____   b. _____   c. _____   d. _____

2. Write the value of each coin.

    a. The value of one dime is _____ cent(s).

    b. The value of one penny is _____ cent(s).

    c. The value of one nickel is _____ cent(s).

    d. The value of one quarter is _____ cent(s).

3. Your mom said she will give you 1 nickel or 1 quarter. Which would you take, and why?

_____

_____

_____

_____

EUREKA MATH™   Lesson 21:   Identify quarters by their image, name, or value. Decompose the value of a quarter using pennies, nickels, and dimes.   85

©2015 Great Minds. eureka-math.org
G1-M5M6-SE-B4-1.3.1-01.2016

4. Lee has 25 cents in his piggy bank. Which coin or coins could be in his bank?

 a. Draw to show the coins that could be in Lee's bank.

 b. Draw a different set of coins that could be in Lee's bank.

          Lesson 21:   Identify quarters by their image, name, or value. Decompose the value
                                     of a quarter using pennies, nickels, and dimes.   **EUREKA
MATH**™

©2015 Great Minds. eureka-math.org
G1-M5M6-SE-B4-1.3.1-01.2016

Name _____   Date _____

1. Use the word bank to label the coins.

   | quarter   dime   nickel   penny |

   a. _____   b. _____   c. _____   d. _____

2. Match the coin combinations to the coin on the right with the same value.

   a.

   •   •

   b.

   •   •

   c.

   •   •

3. Tamra has 25 cents in her hand. Cross off (x) the hand that cannot be Tamra's.

4. Ben thinks he has more money than Peter. Is he correct? Why or why not?

Ben's Money   **Peter's Money**

Ben is _____ because _____

_____

5. Solve. Match each statement to the coin that shows the value of the answer.

   a. 5 pennies = _____ cents   •

   b. 6 cents + 4 cents = _____ cents   •

   c. 1 quarter = _____ cents   •

   d. 6 cents – 5 cents = _____ cent(s)   •

EUREKA MATH™

Name _____   Date _____

1. Match the label to the correct coins, and write the value. There will be more than one match for each coin name.

   a.
   ┌─────────────────────┐
   │   **nickel**   │
   │   │
   │ _____ cents   │
   └─────────────────────┘

   b.
   ┌─────────────────────┐
   │   **dime**   │
   │   │
   │ _____ cents   │
   └─────────────────────┘

   c.
   ┌─────────────────────┐
   │   **quarter**   │
   │   │
   │ _____ cents   │
   └─────────────────────┘

   d.
   ┌─────────────────────┐
   │   **penny**   │
   │   │
   │ _____ cent   │
   └─────────────────────┘

Lesson 22:   Identify varied coins by their image, name, or value. Add one cent to the value of any coin.

89

©2015 Great Minds. eureka-math.org
G1-M5M6-SE-B4-1.3.1-01.2016

2. Lee has one coin in his pocket, and Pedro has 3 coins. Pedro has more money than Lee. Draw a picture to show the coins each boy might have.

Lee's Pocket

Pedro's Pocket

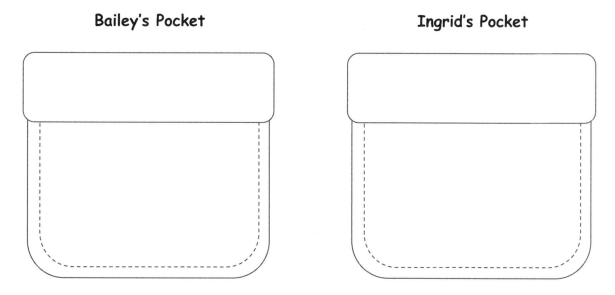

3. Bailey has 4 coins in her pocket, and Ingrid has 4 coins. Ingrid has more money than Bailey. Draw a picture to show the coins each girl might have.

Bailey's Pocket

Ingrid's Pocket

Lesson 22: Identify varied coins by their image, name, or value. Add one cent to the value of any coin.

EUREKA MATH™

Name _____ Date _____

1. Add pennies to show the written amount.

a.		
8 cents		
b.		
30 cents		
c.		
10 cents		
d.		
18 cents		

2. Write the value of each group of coins.

a.

_____ cents

b.

_____ cents

c.

_____ cents

d.

_____ cents

e.

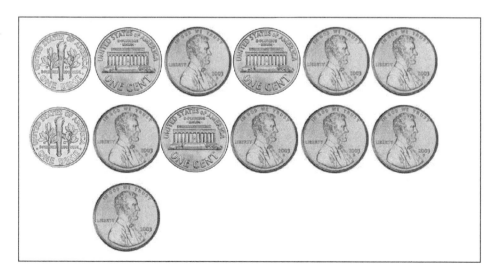

_____ cents

Lesson 23:   Count on using pennies from any single coin.

EUREKA MATH™

Name _____   Date _____

1. Add pennies to show the written amount.

a.	15 cents	
b.	28 cents	
c.	22 cents	
d.	32 cents	

2. Write the value of each group of coins.

a.

_____ cents

b.

_____ cents

c.

_____ cents

d.

_____ cents

e.

_____ cents

Lesson 23:   Count on using pennies from any single coin.

EUREKA
MATH™

Name _____ Date _____

1. Find the value of each set of coins. Complete the place value chart to match. Write an addition sentence to add the value of the dimes and the value of the pennies.

a.

tens	ones

_____

b.

tens	ones

_____

c.

tens	ones

_____

EUREKA MATH

Lesson 24:   Use dimes and pennies as representations of numbers to 120.

95

2. Check the set that shows the correct amount. Fill in the place value chart to match.

a. 80 cents

tens	ones

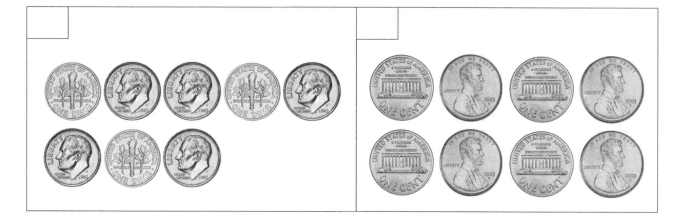

b. 100 cents

tens	ones

3. Draw 58 cents using dimes and pennies. Fill in the place value chart.

tens	ones

Lesson 24:   Use dimes and pennies as representations of numbers to 120.

EUREKA MATH™

Name _____   Date _____

1. Find the value of each set of coins. Complete the place value chart.
   Write an addition sentence to add the value of the dimes and the value of the pennies.

a.

tens	ones

_____

b.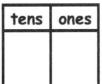

tens	ones

_____

c.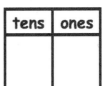

tens	ones

_____

2. Check the set that shows the correct amount. Fill in the place value chart to match.

110 cents

tens	ones

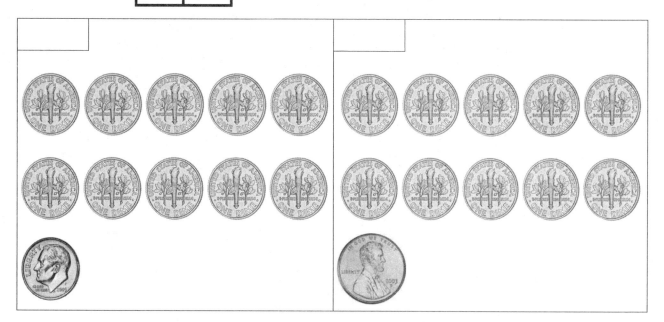

3. a. Draw 79 cents using dimes and pennies. Fill in the place value chart to match.

tens	ones

b. Draw 118 cents using dimes and pennies. Fill in the place value chart to match.

tens	ones

Lesson 24:   Use dimes and pennies as representations of numbers to 120.

EUREKA
MATH

Name _____   Date _____

<u>R</u>ead the word problem.

<u>D</u>raw a tape diagram or double tape diagram and label.

<u>W</u>rite a number sentence and a statement that matches the story.

Sample Tape Diagram

1. Kiana wrote 3 poems. She wrote 7 fewer than her sister Emi. How many poems did Emi write?

_____

2. Maria used 14 beads to make a bracelet. Maria used 4 more beads than Kim. How many beads did Kim use to make her bracelet?

_____

3. Peter drew 19 rocket ships. Rose drew 5 fewer rocket ships than Peter. How many rocket ships did Rose draw?

4. During the summer, Ben watched 9 movies. Lee watched 4 more movies than Ben. How many movies did Lee watch?

_____

5. Anton's family packed 10 suitcases for vacation. Anton's family packed 3 more suitcases than Fatima's family. How many suitcases did Fatima's family pack?

_____

6. Willie painted 9 fewer pictures than Julio. Julio painted 16 pictures. How many pictures did Willie paint?

Name _____   Date _____

Sample Tape Diagram

Read the word problem.
Draw a tape diagram or double tape diagram and label.
Write a number sentence and a statement that matches the story.

1. Julio listened to 7 songs on the radio. Lee listened to 3 more songs than Julio. How many songs did Lee listen to?

2. Shanika caught 14 ladybugs. She caught 4 more ladybugs than Willie. How many ladybugs did Willie catch?

3. Rose packed 3 more boxes than her sister to move to their new house. Her sister packed 11 boxes. How many boxes did Rose pack?

4. Tamra decorated 13 cookies. Tamra decorated 2 fewer cookies than Emi.
    How many cookies did Emi decorate?

5. Rose's brother hit 12 tennis balls. Rose hit 6 fewer tennis balls than her brother.
    How many tennis balls did Rose hit?

6. With his camera, Darnel took 5 more pictures than Kiana. He took 13 pictures.
    How many pictures did Kiana take?

Lesson 25:   Solve compare with bigger or smaller unknown problem types.

EUREKA
MATH

Name _____   Date _____

Read the word problem.
Draw a tape diagram or double tape diagram and label.
Write a number sentence and a statement that matches the story.

Sample Tape Diagram

N [6]
R [6 | 4]
  ?=10
6 + 4 = [10]

1. Tony is reading a book with 16 pages. Maria is reading a book that has 10 pages. How much longer is Tony's book than Maria's book?

_____

2. Shanika built a block tower using 14 blocks. Tamra built a tower by using 5 more blocks than Shanika. How many blocks did Tamra use to build her tower?

_____

3. Darnel walked 10 minutes to get to Kiana's house. The next day, Kiana took a shortcut and walked to Darnel's house in 8 minutes. How much shorter in time was Kiana's walk?

©2015 Great Minds. eureka-math.org
G1-M5M6-SE-B4-1.3.1-01.2016

4. Lee read 16 pages in a book. Kim read 4 fewer pages in her book. How many pages did Kim read?

5. Nikil's soccer team has 13 players. Nikil has 4 fewer players on his team than Rose's team. How many players are on Rose's team?

6. After dinner, Darnel washed 15 spoons. He washed 9 more spoons than forks. How many forks did Darnel wash?

Lesson 26:   Solve compare with bigger or smaller unknown problem types.

EUREKA MATH

©2015 Great Minds. eureka-math.org
G1-M5M6-SE-B4-1.3.1-01.2016

Name _____   Date _____

Read the word problem.
Draw a tape diagram or double tape diagram and label.
Write a number sentence and a statement that matches the story.

Sample tape diagram

N [6]
R [6 | 4]
? = 10
6 + 4 = [10]

1. Fatima walks 15 blocks home from school. Ben walks 8 blocks. How much longer is Fatima's walk home from school than Ben's?

_____

2. Maria bought a basket with 13 strawberries in it. Darnel bought a basket with 4 more strawberries than Maria. How many strawberries did Darnel's basket have in it?

_____

3. Tamra has 5 books checked out from the library. Kim has 11 books checked out from the library. How many fewer books does Tamra have checked out than Kim?

EUREKA MATH

Lesson 26:   Solve compare with bigger or smaller unknown problem types.

105

©2015 Great Minds. eureka-math.org
G1-M5M6-SE-B4-1.3.1-01.2016

4. Kiana picked 12 apples from the tree. She picked 6 fewer apples than Willie. How many apples did Willie pick from the tree?

5. During recess, Emi found 16 rocks. She found 5 more rocks than Peter. How many rocks did Peter find?

6. The first grade football team has 12 players. The first grade team has 6 fewer players than the second grade team. How many players are on the second grade team?

EUREKA
MATH

Name _____ Date _____

Sample Tape Diagram

Read the word problem.
Draw a tape diagram or double tape diagram and label.
Write a number sentence and a statement that matches the story.

N [6]
R [6 | 4]
  ?=10
6 + 4 = [10]

1. Nine letters came in the mail on Monday. Some more letters were delivered on Tuesday. Then, there were 13 letters. How many letters were delivered on Tuesday?

2. Ben and Tamra found a total of 18 seeds in their watermelon slices. Ben found 7 seeds in his slice. How many seeds did Tamra find?

3. Some children were playing on the playground. Eight children came to join, and now there are 14 children. How many children were on the playground in the beginning?

EUREKA
MATH™

Lesson 27:   Share and critique peer strategies for solving problems of varied types.

107

4. Willie walked for 7 minutes. Peter walked for 14 minutes. How much shorter in time was Willie's walk?

---

5. Emi saw 12 ants walking in a row. Fran saw 6 more ants than Emi. How many ants did Fran see?

---

6. Shanika has 13 cents in her front pocket. She has 8 fewer cents in her back pocket. How many cents does Shanika have in her back pocket?

108   Lesson 27:   Share and critique peer strategies for solving problems of varied types.

EUREKA
MATH™

Name _____   Date _____

Sample Tape Diagram

Read the word problem.
Draw a tape diagram or double tape diagram and label.
Write a number sentence and a statement that matches the story.

1. Eight students lined up to go to art. Some more lined up to go to music. Then, there were 12 students in line. How many students lined up to go to music?

_____

2. Peter rode his bike 5 blocks. Rose rode her bike 13 blocks. How much shorter was Peter's ride?

_____

3. Lee and Anton collected 16 leaves on their walk. Nine of the leaves were Lee's. How many leaves were Anton's?

EUREKA MATH

Lesson 27:   Share and critique peer strategies for solving problems of varied types.

109

©2015 Great Minds. eureka-math.org
G1-M5M6-SE-B4-1.3.1-01.2016

4. The team counted 11 soccer balls inside the net. They counted 5 fewer soccer balls outside of the net. How many soccer balls were outside of the net?

_____

5. Julio saw 14 cars drive by his house. Julio saw 6 more cars than Shanika. How many cars did Shanika see?

_____

6. Some students were eating lunch. Four students joined them. Now, there are 17 students eating lunch. How many students were eating lunch in the beginning?

Lesson 27:   Share and critique peer strategies for solving problems of varied types.

EUREKA
MATH™

©2015 Great Minds. eureka-math.org
G1-M5M6-SE-B4-1.3.1-01.2016

Name _____   Date _____

1. Circle the smiley face that shows your level of fluency for each activity.

Activity	I still need some practice.	I can complete, but I still have some questions.	I am fluent.
a.			
b.			
c.			
d.			
e.			
f.			

2. Which activity helped you the most in becoming fluent with your facts to 10?

EUREKA MATH™   **Lesson 28:**   Celebrate progress in fluency with adding and subtracting within 10 (and 20). Organize engaging summer practice.   **111**

©2015 Great Minds. eureka-math.org
G1-M5M6-SE-B4-1.3.1-01.2016

Name _____ Date _____

1. Teach a family member some of our counting activities. Check all the activities you do together.

   ☐ Happy Count by ones.
   ☐ Happy Count by tens.
   ☐ Count by ones the Say Ten Way.
   ☐ Count by tens the Say Ten Way. First, start at 0; then, start at 7.
   ☐ Movement counting—count while doing squats, arm rolls, jumping jacks, etc.

2. Write the numbers from 91 to 120:

91		93							

				105					

								119	

3. Count backward by tens from 97 to 7.

   97, _____, 77, _____, _____, _____, _____, _____, _____, _____

4. On the back of your paper, write as many sums and differences within 20 as you can. Circle the ones that were hard for you at the beginning of the year!

Lesson 28: Celebrate progress in fluency with adding and subtracting within 10 (and 20). Organize engaging summer practice.

EUREKA MATH

Name _____ Date _____

Complete a math activity each day. Color the box for each day you do the suggested activity.

Summer Math Review: Weeks 1–5

	Monday	**Tuesday**	**Wednesday**	**Thursday**	**Friday**
Week 1	Count from 87 to 120 and back.	Play Addition with Cards.	Use your tangram pieces to make a Fourth of July picture.	Use quick tens and ones to draw 76.	Complete a Sprint.
Week 2	Do counting squats. Count from 45 to 60 and back the Say Ten Way.	Play Subtraction with Cards.	Make a graph of the types of fruits in your kitchen. What did you find out from your graph?	Solve 36 + 57. Draw a picture to show your thinking.	Complete a Sprint.
Week 3	Write numbers from 37 to as high as you can in one minute, while whisper-counting the Say Ten Way.	Play Target Practice or Shake Those Disks for 9 and 10.	Measure a table with spoons and then with forks. Which did you need more of? Why?	Use real coins or draw coins to show as many ways to make 25 cents as you can.	Complete a Sprint.
Week 4	Do jumping jacks as you count up by tens to 120 and back down to 0.	Play Race and Roll Addition or Addition with Cards.	Go on a shape scavenger hunt. Find as many rectangles or rectangular prisms as you can.	Use quick tens and ones to draw 45 and 54. Circle the greater number.	Complete a Sprint.
Week 5	Write the numbers from 75 to 120.	Play Race and Roll Subtraction or Subtraction with Cards.	Measure the route from your bathroom to your bedroom. Walk heel to toe, and count your steps.	Add 5 tens to 23. Add 2. What number did you find?	Complete a Sprint.

Name _____   Date _____

Complete a math activity each day. Color the box for each day you do the suggested activity.

Summer Math Review: Weeks 6–10

	Monday	Tuesday	Wednesday	Thursday	Friday
Week 6	Count by ones from 112 to 82. Then, count from 82 to 112.	Play Missing Part for 7.	Write a story problem for 9 + 4.	Solve 64 + 38. Draw a picture to show your thinking.	Complete a Core Fluency Practice Set.
Week 7	Do counting squats. Count down from 99 to 75 and back up the Say Ten Way.	Play Race and Roll Addition or Addition with Cards.	Graph the colors of all your pants. What did you find out from your graph?	Draw 14 cents with dimes and pennies. Draw 10 more cents. What coins did you use?	Complete a Core Fluency Practice Set.
Week 8	Write the numbers from 116 to as low as you can in one minute.	Play Missing Part for 8.	Write a story problem for 7 + ___ = 12.	Use quick tens and ones to draw 76. Draw dimes and pennies to show 59 cents.	Complete a Core Fluency Practice Set.
Week 9	Do jumping jacks as you count up by tens from 9 to 119 and back down to 9.	Play Race and Roll Subtraction or Subtraction with Cards.	Go on a shape scavenger hunt. Find as many circles or spheres as you can.	Use quick tens and ones to draw 89 and 84. Circle the number that is less.	Complete a Core Fluency Practice Set.
Week 10	Write numbers from 82 to as high as you can in one minute, while whisper counting the Say Ten Way.	Play Target Practice or Shake Those Disks for 6 and 7.	Measure the steps from your bedroom to the kitchen, walking heel to toe, and then have a family member do the same thing. Compare.	Solve 47 + 24. Draw a picture to show your thinking.	Complete a Core Fluency Practice Set.

Lesson 30:   Create folder covers for work to be taken home illustrating the year's learning.

EUREKA MATH™

Addition (or Subtraction) with Cards

Materials: 2 sets of numeral cards 0–10

- Shuffle the cards, and place them face down between the two players.
- Each partner flips over two cards and adds them together or subtracts the smaller number from the larger one.
- The partner with the largest sum or smallest difference keeps the cards played by both players in that round.
- If the sums or differences are equal, the cards are set aside, and the winner of the next round keeps the cards from both rounds.
- When all the cards have been used, the player with the most cards wins.

Sprint

Materials: Sprint (Sides A and B)

- Do as many problems on Side A as you can in one minute. Then, try to see if you can improve your score by answering even more of the problems on Side B in a minute.

Target Practice

Materials: 1 die

- Choose a target number to practice (e.g., 10).
- Roll the die, and say the other number needed to hit the target. For example, if you roll 6, say 4, because 6 and 4 make ten.

Shake Those Disks

Materials: Pennies

The amount of pennies needed depends on the number being practiced. For example, if students are practicing sums for 10, they need 10 pennies.

- Shake your pennies, and drop them on the table.
- Say two addition sentences that add together the heads and tails. (For example, if they see 7 heads and 3 tails, they would say 7 + 3 = 10 and 3 + 7 = 10.)
- Challenge: Say four addition sentences instead of two. (For example, 10 = 7 + 3, 10 = 3 + 7, 7 + 3 = 10, and 3 + 7 = 10.)

       Lesson 30:   Create folder covers for work to be taken home illustrating the year's   **115**
learning.

©2015 Great Minds. eureka-math.org
G1-M5M6-SE-B4-1.3.1-01.2016

Race and Roll Addition (or Subtraction)

Materials: 1 die

Addition

- Both players start at 0.
- They each roll a die and then say a number sentence adding the number rolled to their total. (For example, if a player's first roll is 5, the player says 0 + 5 = 5.)
- They continue rapidly rolling and saying number sentences until someone gets to 20 without going over. (For example, if a player is at 18 and rolls 5, the player would continue rolling until she gets a 2.)
- The first player to 20 wins.

Subtraction

- Both players start at 20.
- They each roll a die and then say a number sentence subtracting the number rolled from their total. (For example, if a player's first roll is 5, the player says 20 – 5 = 15.)
- They continue rapidly rolling and saying number sentences until someone gets to 0 without going over. (For example, if a player is at 5 and rolls 6, the player would continue rolling until she gets a 5.)
- The first player to 0 wins.

Missing Part: Make 7 or 8

Materials: 5-Group Cards

- Each partner holds a card up to his or her forehead.
- One partner tells how many more are needed to make 7 (or 8).
- The other partner must guess the cards on his or her forehead.
- Partners can play simultaneously, each putting a card to his or her forehead.

Lesson 30:   Create folder covers for work to be taken home illustrating the year's learning.

©2015 Great Minds. eureka-math.org
G1-M5M6-SE-B4-1.3.1-01.2016

EUREKA MATH™